T0256872

Information and Instructions

This shop manual contains several sections each covering a specific group of wheel type tractors. The Tab Index on the preceding page can be used to locate the section pertaining to each group of tractors. Each section contains the necessary specifications and the brief but terse procedural data needed by a mechanic when repairing a tractor on which he has had no previous actual experience.

Within each section, the material is arranged in a systematic order beginning with an index which is followed immediately by a Table of Condensed Service Specifications. These specifications include dimensions, fits, clearances and timing instructions. Next in order of arrangement is the procedures paragraphs.

In the procedures paragraphs, the order of presentation starts with the front axle system and steering and proceeding toward the rear axle. The last paragraphs are devoted to the power take-off and power lift systems. Interspersed where needed are additional tabular specifications pertaining to wear limits, torquing, etc.

HOW TO USE THE INDEX

Suppose you want to know the procedure for R&R (remove and reinstall) of the engine camshaft. Your first step is to look in the index under the main heading of ENGINE until you find the entry "Camshaft." Now read to the right where under the column covering the tractor you are repairing, you will find a number which indicates the beginning paragraph pertaining to the camshaft. To locate this wanted paragraph in the manual, turn the pages until the running index appearing on the top outside corner of each page contains the number you are seeking. In this paragraph you will find the information concerning the removal of the camshaft.

More information available at haynes.com
Phone: 805-498-6703

Haynes Group Limited
Haynes North America, Inc.

ISBN-10: 0-87288-102-4
ISBN-13: 978-0-87288-102-0

© **Haynes North America, Inc. 1991**
With permission from Haynes Group Limited

Clymer is a registered trademark of Haynes North America, Inc.

Cover art by Sean Keenan

All rights reserved. No part of this book may be reproduced or transmitted in any form or by any means, electronic or mechanical, including photocopying, recording or by any information storage or retrieval system, without permission in writing from the copyright holder.

While every attempt is made to ensure that the information in this manual is correct, no liability can be accepted by the authors or publishers for loss, damage or injury caused by any errors in, or omissions from, the information given.

IH-10, 6T1, 15-80

SHOP MANUAL
INTERNATIONAL HARVESTER

MODELS COVERED AND IDENTIFICATION

TRACTOR SERIES	COMPLETE TRACTOR MODEL DESIGNATION	TRACTOR MODEL (ABBREVIATED)	ENGINE MODEL	VERSIONS BUILT
Series 300	Farmall 300	300	C169	2, 3, 4
	Farmall 300 Hi-Clear	300HC	C169	2
Series 300U	International 300 Utility	I 300U	C169	2
Series 350	Farmall 350	350	C175	2, 3, 4
	Farmall 350 Hi-Clear	350HC	C175	2
Series 350U	International 350 Utility	I 350U	C175	2
Series 350D	Farmall 350 Diesel	350D	D193	2, 3, 4
	Farmall 350 Diesel Hi-Clear	350DHC	D193	2
Series 350DU	International 350 Diesel Utility	I350DU	D193	2
Series 400	Farmall 400	400	C264	2, 3, 4
	Farmall 400 Hi-Clear	400HC	C264	2
Series 400D	Farmall 400 Diesel	400D	D264	2, 3, 4
	Farmall 400 Diesel Hi-Clear	400DHC	D264	2
Series W400	International W400	W 400	C264	1
Series W400D	International W400 Diesel	W 400D	D264	1
Series 450	Farmall 450	450	C281	2, 3, 4
	Farmall 450 Hi-Clear	450HC	C281	2
Series 450D	Farmall 450 Diesel	450D	D281	2, 3, 4
	Farmall 450 Diesel Hi-Clear	450DHC	D281	2
Series W450	International W450	W 450	C281	1
Series W450D	International W450 Diesel	W 450D	D281	1

1. Non-Adjustable Front Axle
2. Adjustable Front Axle
3. Dual Wheel Tricycle
4. Single Wheel Tricycle

Engine Serial Number is located on side of crankcase.

Tractor Serial Number is located on side of clutch housing.

Suffix letters to the Serial Number Indicate the following attachments:

A. Distillate Burning
B. Kerosene Burning
C. LP-Gas Burning (Standard Altitude)
D. 5000-Foot Altitude
E. 8000-Foot Altitude
F. Cotton Picker Mounting (High Drum)
G. Cotton Picker Tractor Attachment (High Drum)

H. Rear Frame Cover and Gear Shifter
J. Rockford Clutch
N. LP-Gas Burning (2500-Foot Altitude and Up)
P. Independent PTO Drive for Use Without Torque Amplifier
R. Torque Amplifier With Provision for Transmission Driven PTO
S. Torque Amplifier With Provision for Independent PTO
T. Cotton Picker Mounting (Low Drum)

INDEX (By Starting Paragraph)

CONDENSED SERVICE DATA

	300-300HC	300 Utility	Non-Diesel 350-350HC	Non-Diesel 350 Utility	Non-Diesel 400-400HC-W400	Diesel 400-400HC-W400	Non-Diesel 450-450HC-W450	Diesel 450-450HC-W450
GENERAL								
Engine Make	Own	Own	Own	Own	Own	Own	Own	Own
Engine Model	C-169	C-169	C-175	C-175	C-264	D-264	C-281	D-281
Number of Cylinders	4	4	4	4	4	4	4	4
Bore—Inches	3 9/16	3 9/16	3 5/8	3 5/8	4	4	4 1/8	4 1/8
Stroke—Inches	4 1/4	4 1/4	4 1/4	4 1/4	5 1/4	5 1/4	5 1/4	5 1/4
Displacement—Cubic Inches	169	169	175	175	264	264	281	281
Pistons Removed From	Above	Above	Above	Above	Above	Above	Above	Above
Main Bearings, Number of	3	3	3	3	3	5	3	5
Main and Rod Bearings Adjustable?	No	No	No	No	No	No	No	No
Cylinder Sleeves	Dry	Dry	Dry	Dry	Dry	Dry	Dry	Dry
Forward Speeds, No T.A.	5	5	5	5	5	5	5	5
Forward Speeds, With T.A.	10	10	10	10	10	10	10	10
Generator and Starter Make	D-R	D-R	D-R	D-R	D-R	D-R	D-R	D-R
TUNE-UP								
Firing Order	1-3-4-2	1-3-4-2	1-3-4-2	1-3-4-2	1-3-4-2	1-3-4-2	1-3-4-2	1-3-4-2
Valve Tappet Gap	0.017H	0.017H	0.017H	0.017H	0.017H	0.017H	0.017H	0.017H
Inlet Valve Seat Angle	45	45	45	45	45	45	45	45
Exhaust Valve Seat Angle	45	45	45	45	45	45	45	45
Ignition Distributor Make	Own	Own	Own	Own	Own	Own	Own	Own
Ignition Distributor Symbol—								
Distillate or Kerosene	J	J	J	J	J	....	J	....
Gasoline or Diesel	O	S	O	S	N	H	N	H or C
LP-Gas	T	T	T	T	T	....	T	....
Ignition Magneto Make	....	Own	....	Own	....	....	....	....
Ignition Magneto Model	....	H-4	....	H-4	....	....	....	....
Breaker Gap, Distributor	0.020	0.020	0.020	0.020	0.020	0.020	0.020	0.020
Breaker Gap, Magneto	....	0.013	....	0.013	....	....	....	....
Distributor Timing, Retard	TDC	TDC	TDC	TDC	TDC	TDC	TDC	TDC
Distributor Timing, Advanced—								
Distillate or Kerosene	30°B	30°B	30°B	30°B	30°B	....	30°B	....
Gasoline	22°B	22°B	22°B	22°B	25°B	....	25°B	....
LP-Gas	16°B	16°B	16°B	16°B	16°B	....	16°B	....
Magneto Impulse Trip Point	....	TDC	....	TDC	....	....	....	....
Pulley Mark Indicating—								
Ignition Retard Timing				Last Notch				
Spark Plug Electrode Gap	0.023	0.023	0.023	0.023	0.023	0.023	0.023	0.023
Carburetor Make, LP-Gas	Ensign	Ensign	Ensign	Ensign	Ensign	....	Ensign	....
Carburetor Make, Except LP-Gas	I-H	I-H	I-H	I-H	I-H	I-H	I-H	I-H
Carburetor Model, Except LP-Gas	1 1/4	1 1/4	1 1/4	1 1/4	1 1/4	F8	1 1/4	F8
Carburetor Fuel Level	9/16	9/16	9/16	9/16	9/16	13/32	9/16	13/32
Carburetor Calibration				Refer to Parts Catalogs				
Battery Terminal Grounded	Positive	Positive	Positive	Positive	Positive	Positive	Positive	Positive
Engine Low Idle rpm	425	425	425**	425**	425**	625	425**	625
Engine High Idle rpm, No Load	1925	2200	1925	2200	1600	1580	1600	1580
Engine Full Load rpm	1750	2000	1750	2000	1450	1450	1450	1450
SIZES—CAPACITIES—CLEARANCES								
(Clearances in Thousands)								
Crankshaft Main Journal Diameter	2.558	2.558	2.558	2.558	2.808	3.748	2.808	3.748
Crankpin Diameter	2.298	2.298	2.298	2.298	2.548	3.248	2.548	3.248
Camshaft Journal Diameter, No. 1 (Front)	1.931	1.931	1.931	1.931	2.2435	2.431	2.2435	2.431
Camshaft Journal Diameter, No. 2	1.806	1.806	1.806	1.806	2.1185	2.306	2.1185	2.306
Camshaft Journal Diameter, No. 3	1.3685	1.3685	1.3685	1.3685	1.8685	2.181	1.8685	2.181
Camshaft Journal Diameter, No. 4	....	....	....	....	....	1.8685	....	1.8685
Piston Pin Diameter	††	††	0.87485	0.87485	††	1.3126	0.99985	1.3126
Valve Stem Diameter	0.341	0.341	0.341	0.341	0.3715		0.3715	0.3715
Diesel Starting Valve Stem Diameter	....					0.309		0.309
Main Bearings, Diameter Clearance	1.1-3.7	1.1-3.7	1.1-3.7	1.1-3.7	2.0-3.0	1.8-4.8	2.0-3.0	1.8-4.8
Rod Bearings, Diameter Clearance	1.1-3.7	1.1-3.7	1.1-3.7	1.1-3.7	1.1-3.7	2.0-3.0	1.1-3.7	2.0-3.0
Piston Skirt Clearance	‡‡	‡‡	2.0-3.0	2.0-3.0	‡‡	4.6-5.4	2.5-3.5	4.6-5.4
Crankshaft End Play	4.0-8.0	4.0-8.0	4.0-8.0	4.0-8.0	4.0-8.0	4.0-8.0	4.0-8.0	4.0-8.0
Camshaft Bearings, Diameter Clearance	1.5-3.5	1.5-3.5	1.5-3.5	1.5-3.5	1.5-3.5	1.5-3.5	1.5-3.5	1.5-3.5
Cooling System—Gallons	4 3/8	4	4 7/8	4 1/2	6 1/2*	6 3/4§	6 7/8†	7 5/8‡
Crankcase Oil—Quarts	6	6	6	6	8	9	8	9
Transmission and Differential—Quarts	28	28	28	28	60	60	60	60
Final Drive, Each—Quarts (High Clearance)	3	....	3	....	3	3	3	3
Live PTO Housing (Rear Unit)—Quarts	2	2	2	2	2	2	2	2
TIGHTENING TORQUES—FT.-LBS.								
Cylinder Head	70	70	70	70	110	110-135	110	110-135
Rod Bolts	40	40	40	40	55-60	115	55-60	115
Main Bearing Bolts, Center	75	75	75	75	100-105	250-275	100-105	250-275
Main Bearing Bolts, Others	75	75	75	75	100-105	150-175	100-105	150-175

*Series W400, 6¾ gallons.
†Series W450, 7 gallons.
‡Series W450D, 7¾ gallons.
§Series W400D, 7¼ gallons.

**For LP-Gas tractors, refer to paragraph 72.
††Refer to paragraph 61A.
‡‡Refer to paragraph 54.

CONDENSED SERVICE DATA

GENERAL

	Diesel 350-350HC	Diesel 350 Utility
Engine Make	Continental	Continental
Engine Model	D-193	D-193
Number of Cylinders	4	4
Bore—Inches	3¾	3¾
Stroke—Inches	4⅜	4⅜
Displacement—Cubic Inches	193	193
Pistons Removed From	Above	Above
Main Bearings, Number of	3	3
Main and Rod Bearings Adjustable?	No	No
Cylinder Sleeves	Dry	Dry
Forward Speeds, No T.A.	5	5
Forward Speeds, With T.A.	10	10
Generator and Starter Make	D-R	D-R

TUNE-UP

	Diesel 350-350HC	Diesel 350 Utility
Firing Order	1-3-4-2	1-3-4-2
Valve Tappet Gap	0.014H	0.014H
Inlet Valve Seat Angle	45	45
Exhaust Valve Seat Angle	45	45
Battery Terminal Grounded	Positive	Positive
Engine Low Idle rpm	600	600
Engine High Idle rpm, No Load	1925	2200
Engine Full Load rpm	1750	2000

SIZES—CAPACITIES—CLEARANCES
(Clearances in Thousandths)

	Diesel 350-350HC	Diesel 350 Utility
Crankshaft Main Journal Diameter	2.3745	2.3745
Crankpin Diameter	2.2495	2.2495
Camshaft Journal Diameter—		
No. 1 (Front)	1.8085	1.8085
No. 2	1.746	1.746
No. 3	1.6835	1.6835
Piston Pin Diameter	1.1092	1.1092
Intake Valve Stem Diameter	0.341	0.341
Exhaust Valve Stem Diameter	0.3386	0.3386
Main Bearings, Diameter Clearance	0.9-3.6	0.9-3.6
Rod Bearings, Diameter Clearance	0.6-3.1	0.6-3.1
Piston Skirt Clearance	4.5-5.5	4.5-5.5
Crankshaft End Play	4.0-8.0	4.0-8.0
Camshaft Bearings, Diameter Clearance	3.0-6.0	3.0-6.0
Cooling System—Gallons	4.31	4.0
Crankcase Oil—Quarts	5	5
Transmission and Differential—Quarts	28	28
Final Drive, Each—Quarts (High Clearance)	3	....
Live PTO Housing (Rear Unit)—Quarts	2	2

FRONT SYSTEM TRICYCLE TYPE

Series 300-350-350D-400-400D-450-450D

1. Dual front wheels are mounted on a horizontal axle and lower bolster assembly as shown in Fig. IH600. The single front wheel is mounted in a wheel fork as shown in Fig. IH601. In either case, the wheel fork or lower bolster is attached to the lower portion of the upper bolster pivot shaft by three cap screws and one steering stop bolt (1).

2. Two types of wheel and axle assemblies (Fig. IH601) are available for the fork mounted single front wheel. One has a conventional wheel (13), solid hub (12) and axle shaft (6) with tapered roller bearings (8 & 10). The other type wheel (21) consists of two halves (male and female) and axle shaft (29) with bushings (20). The procedure for removing either type is evident after an examination of the unit.

1. Stop bolt
2. Lower bolster
3. Dust shield
5. Felt washer
6. Oil seal
7. Bearing cone
8. Felt washer retainer
9. Bearing cup
10. Grease retainer
11. Wheel
12. Grease retainer
13. Bearing cup
14. Bearing cone
15. Washer
16. Nut
17. Gasket
18. Hub cap

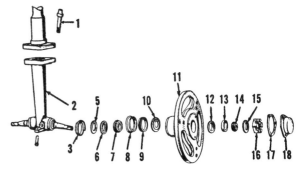

Fig. IH600—Dual wheel tricycle lower bolster, horizontal axle and wheel components.

FRONT SYSTEM—AXLE TYPE

Fig. IH601—Exploded view shows both types of fork mounted single front wheels. The conventional axle (6) is carried in taper roller bearings; whereas axle (29) is carried in two bushings (20).

1. Stop bolt	12. Hub assembly
2. Wheel fork	13. Wheel
3. Nut	20. Bushing
4. Lock washer	21. Wheel halves
5. Dust shield	23. Nut
6. Axle	24. Lock washer
7. Oil seal	25. Shield
8. Bearing cone	26. Felt washer
9. Oil seal retainer	27. Oil seal
10. Bearing cup	28. Grease fitting
11. Grease retainer	29. Axle

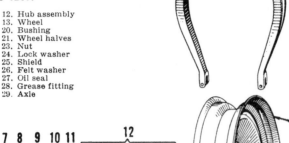
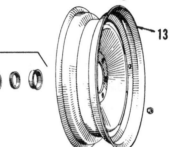
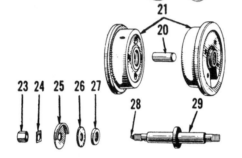

AXLE MAIN MEMBER

Series 300-350-350D-400-400D-W400-W400D-450-450D-W450-W450D

3. On adjustable axle models shown in Figs. IH602 & 602A, the center member pivots on pin (21) which is pinned in the adapter or lower bolster (19). Two pivot pin bushings (12) which are pressed into the center member, should be reamed after installation to provide a suggested pin to bushing clearance of 0.003-0.006.

On non-adjustable axle models, the front axle (21—Fig. IH603) pivots on pin (15) which is carried in bushings (22). After installing new bushings, ream them, if necessary, to provide a free fit for pivot pin (15).

Series 300U-350U-350DU

4. To remove the axle, wheels and knuckles as an assembly, unbolt the radius rod ball bracket (11—Fig. IH-604) from clutch housing and disconnect drag links from knuckle arms. Raise front of tractor just enough to take weight off front wheels and remove the grease fitting from front end of pivot pin (14). Unbolt pivot bracket from bolster (18), continue raising front of tractor and roll the axle, pivot bracket and wheels as-

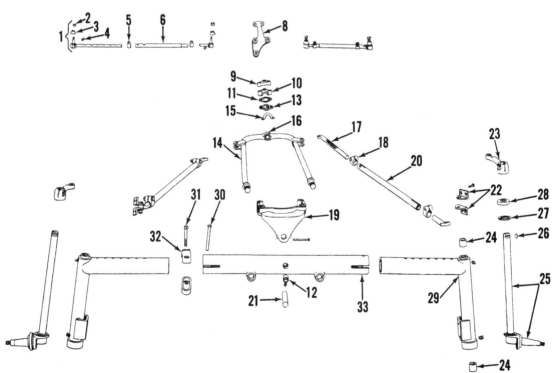

Fig. IH602—Adjustable axle and components used on "Hi-Clear" models. Front wheel toe-in should be ⅛ to ¼-inch. Refer to legend under Fig. IH602A.

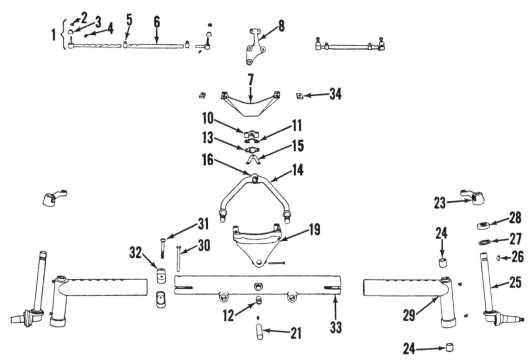

Fig. IH602A—Exploded view of the special adjustable front axle available for series 300, 350, 350D, 400, 400D, 450 and 450D.

1. Tie rod assembly	7. Ball socket support	14. Stay rod	24. Bushings	30. Clamp pin
2. Nut	8. Steering gear arm	15. Lock plate	25. Steering knuckle	31. Clamp bolt
3. Dust cover	10. Ball socket	16. Stay rod ball	26. Woodruff key	32. Axle clamp
4. Fitting	11. Shim	19. Lower bolster	27. Felt washer	33. Axle center member
5. Clamp	12. Bushings	21. Axle pivot pin	28. Thrust bearing	34. Washer
6. Tie rod tube	13. Ball socket cap	23. Steering knuckle arm	29. Axle extension	

sembly forward and away from tractor. Remove the cap screw retaining pivot pin to pivot bracket and bump the pivot pin rearward and out of pivot bracket.

Axle center member (16) is fitted with two bushings (20) which have an inside diameter of 1.251-1.255. Replacement bushings are pre-sized and will not require reaming if not distorted during installation. Install the bushings flush with front and rear of axle bore. The 1.233-1.234 diameter pivot pin (14) has a clearance of 0.017-0.022 in the bushings.

Spherical diameter of the renewable radius rod pivot ball is 1.368-1.378. On 300 Utility models prior to Serial No. 26924, the radius rod ball shank was pressed and riveted into the radius rod. To remove the ball, grind-off the peened portion and press shank from radius rod bore. On later models, the pivot ball unit is also pressed into the radius rod bore, but the shank is threaded for a retaining nut which should be tightened to a torque of 110 Ft.-Lbs. The late production unit can be used for service installation on earlier models.

When reassembling, be sure the side of axle marked "TOP" (stamped above pivot pin location) is up. Vary the number of shims (12) to provide a

snug fit for the radius rod pivot ball without binding. On models where the ball bracket is retained with ½-inch cap screws tighten screws to a torque of 120 Ft.-Lbs. Tighten ⅝-inch screws to a torque of 140-160 Ft.-Lbs.

When assembly is complete jack up front of tractor, oscillate axle to both extreme positions and check the point of contact between the axle and bolster support bracket recess. If the point of contact is not in the center of

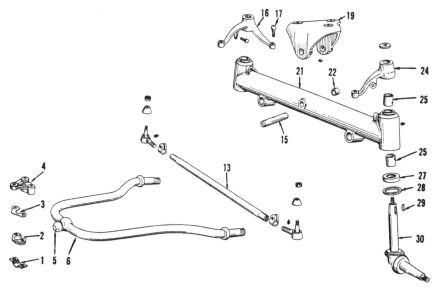

Fig. IH603—Front axle and associated parts used on models W400, W400D, W450 and W450D. Recommended front wheel toe-in is ¼ to ⅜-inch.

1. Lock plate	13. Tie rod	24. Right steering arm
2. Ball cap	15. Axle pivot pin	25. Bushings
3. Shim	16. Left steering arm	27. Thrust bearing
4. Ball socket	17. Ball stud	28. Felt washer
5. Stay rod ball	19. Bolster	29. Woodruff key
6. Stay rod	21. Axle	30. Knuckle
	22. Bushings	

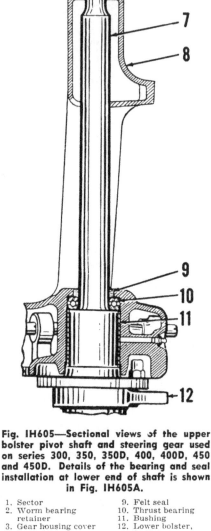

the recess as shown by the arrow in Fig. IH604A, grind-out the recess enough to provide free movement of the axle to both extreme positions.

TIE RODS AND DRAG LINK
All Models So Equipped

5. Procedure for removing the tie rod or rods, tie rod tubes, tie rod ends and/or drag links is self-evident after an examination of the unit and reference to Figs. IH602, 602A, 603 and 604.

Adjust the toe-in of the front wheels to $\frac{1}{8}$-$\frac{1}{4}$-inch on adjustable axle types and $\frac{1}{4}$-$\frac{3}{8}$-inch on non-adjustable type axles. Adjustment is made by varying the length of the tie rods or drag links. On Utility models, alignment marks are provided on the steering arms & axle extensions as well as on the Pitman arms and gear housing to facilitate making the toe-in adjustment. Drag links are properly adjusted if all marks align at the same time.

STEERING KNUCKLES
All Models So Equipped

6. Procedure for removing knuckles from axle or axle extensions is evident

after an examination of the unit and reference to Figs. IH602, 602A, 603 or 604.

When installing new knuckle bushings, be sure to align oil hole in bushing with oil hole in axle or axle extension. Ream the bushings after installation, if necessary, to provide a suggested minimum clearance of 0.003 for the knuckles.

Fig. IH604A—On some 300 Utility models it may be necessary to grind out the bolster support bracket recess to provide free movement of the axle to both extreme positions. Axle should contact the recess at point indicated by the arrow.

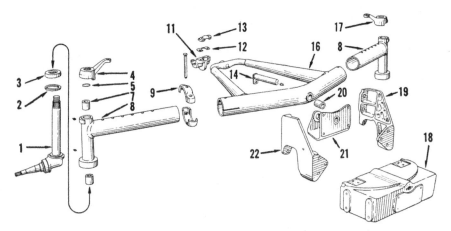

Fig. IH604—Models 300U, 350U and 350DU adjustable front axle, pivot bracket and associated parts.

1. Knuckle
2. Felt washer
3. Thrust bearing
4. Right steering arm
5. Snap ring
7. Bushings
8. Axle extensions
9. Axle clamp
11. Ball bracket
12. Shims
13. Ball cap
14. Pivot pin
16. Front axle
17. Left steering arm
18. Bolster
19. Left bolster support
20. Bushing
21. Pivot bracket
22. Right bolster support

Fig. IH605—Sectional views of the upper bolster pivot shaft and steering gear used on series 300, 350, 350D, 400, 400D, 450 and 450D. Details of the bearing and seal installation at lower end of shaft is shown in Fig. IH605A.

1. Sector
2. Worm bearing retainer
3. Gear housing cover
4. Sector nut
5. Bushing
6. Oil seal
7. Upper bolster pivot shaft
8. Upper bolster
9. Felt seal
10. Thrust bearing
11. Bushing
12. Lower bolster, fork or arm
13. Bearing
14. Bushing
15. Felt seal
16. Worm and shaft

MANUAL STEERING SYSTEM

Series 300-350-350D-400-400D-450-450D

The non-adjustable worm and sector type steering gear is contained in the upper portion of the upper bolster as shown in Fig. IH605.

7. **REMOVE AND REINSTALL.** To remove the steering gear, bolster,

front wheels, grille and radiator as an assembly, proceed as follows: Remove hood and on models with manual steering, disconnect the steering shaft universal joint and remove the U-joint Woodruff key. On models with power steering, loosen the steering shaft coupling which is located in front

of the power unit. Disconnect radiator hoses and remove dust pan from under front of frame rails. Unbolt the radiator brace from water outlet casting; or, on models so equipped, remove the radiator upper support rod.

Disconnect the shutter control rod and on models with an adjustable type front axle, disconnect the radius rod at its rear ball joint. Support tractor under clutch housing and remove cap screws which retain the upper bolster to the side rails. Move front end assembly forward and away from tractor.

8. **OVERHAUL.** The steering gear unit can be overhauled without removing the bolster unit from the tractor. To remove the steering worm and shaft, remove hood and grille, disconnect the steering shaft universal joint and remove the U-joint Woodruff key. Remove the worm shaft bearing retainer (2—Fig. IH605) and withdraw shaft by rotating same forward and out of the steering gear housing. Steering worm shaft bushing (14) which is pre-sized, and oil seal (15) can be renewed at this time.

To remove the sector, remove hood, grille, steering gear housing cover (3) and wormshaft. Remove sector retaining nut (4) and using a suitable puller, remove sector from upper bolster pivot shaft.

Renewal of upper bolster pivot shaft oil seal (6), bushings (5) and (11) and/or thrust bearing (10), requires removal of the upper bolster pivot shaft as follows:

To remove the upper bolster pivot shaft after worm shaft and sector are removed, support front of tractor and remove front axle, lower bolster or wheel fork. Withdraw the pivot shaft from below. The pre-sized bushings (5 & 11) and oil seal (6) can be renewed at this time. End play of upper bolster pivot shaft is non-adjustable; vertical thrust being taken on ball thrust bearing (10). Install the ball thrust bearing so that race with smaller bore is toward bottom of bolster as

shown in Fig. IH605A. The spacer shown is used on series 300, 350 and 350D only.

Series W400-W400D-W450-W450D

9. **ADJUSTMENT.** The steering gear is of the cam and lever type mounted on rear frame (transmission case) cover. Camshaft end play and gear backlash are adjustable, as follows:

10. **ADJUST CAMSHAFT END PLAY.** This adjustment is controlled by shims (10—Fig. IH606) located between steering tube (9) and steering gear housing (17). Adjustment is correct when cam (16) has zero end play and yet turns freely. To decrease end play remove shims. Shims are available in thicknesses of 0.002, 0.003 and 0.010.

11. **ADJUST BACKLASH.** The drag link must be disconnected before making this adjustment. After adjusting the steering cam shaft end play, as in paragraph 10, disconnect the drag link and place gear on the high point by turning steering wheel to mid-position of its rotation. Tighten cross shaft adjusting screw (30) until a very slight drag is felt only at the mid-point when turning steering wheel slowly from the full right to the full left position. Gear should rotate freely at all other points.

12. **R&R AND OVERHAUL.** To remove the steering gear housing assembly, first disconnect engine speed

control linkage and Hydra-Touch control handle bracket from steering column. Disconnect the drag link from Pitman arm. Remove the gear housing shields. Remove the cap screws retaining gear housing to transmission housing cover and remove housing assembly from tractor.

To disassemble the unit, correlation mark the Pitman arm with respect to the lever shaft and remove the Pitman arm. Remove the housing cover (29) and withdraw cam lever and shaft. Remove the cap screws retaining steering column (9) to housing (17) and withdraw steering column, cam and bearing as a unit.

The cam shaft lever stud assembly (24) is renewable.

When reassembling the gear unit, reverse the disassembly procedure and when installing the Pitman arm, be sure to align the previously affixed correlation marks.

Series 300U-350U-350DU

13. **ADJUSTMENT.** The steering gear unit is provided with four adjustments: The cam shaft end play, the mesh position between the lever shaft stud and cam, and the end play of the gear shaft can be adjusted without removing the steering gear unit from tractor. The mesh position between the lever shaft and gear shaft gear teeth is adjustable; however, this adjustment requires reaming and installation of oversize dowels and

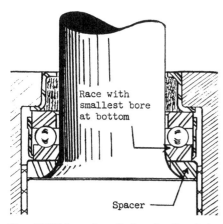

Fig. IH605A — Installation details of the upper bolster pivot shaft lower bearing, seal, etc., used on series 300, 350 and 350D. Series 400, 400D, 450 and 450D are similar except the spacer shown is not used.

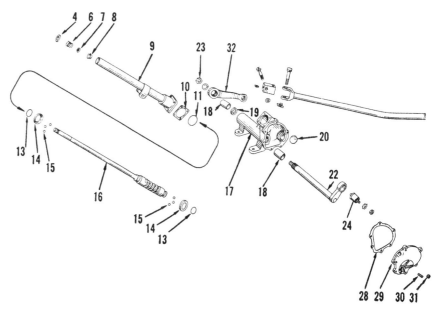

Fig. IH606—Exploded view of the steering gear unit used on series W400, W400D, W450 and W450D. Camshaft end play and backlash are adjustable.

4. Dust seal	11. Seal ring	23. Nut
6. Spring	13. Retaining ring	24. Stud
7. Spring seat	14. Ball cup	28. Gasket
8. Bearing	15. Bearing balls	29. Side cover
9. Jacket tube	16. Cam shaft	30. Adjusting screw
10. Shims (0.002, 0.003	17. Gear housing	31. Adjusting nut
and 0.010)	18. Bushings	32. Pitman arm
	19. Oil seal	
	20. Expansion plug	
	22. Lever shaft	

should not be done except during overhaul and when a new side cover, gear housing, lever shaft and/or gear shaft are installed.

Before attempting to make any adjustments, first make certain that the gear housing is properly filled with lubricant, then disconnect the drag links from the steering (Pitman) arms.

13A. CAMSHAFT END PLAY. To check and/or adjust the steering cam shaft end play, loosen the lock nuts (8—Fig. IH606A) and back-off the adjusting screws (9) three or four turns. Pull up and push down on the steering wheel to detect any end play of the camshaft (27). Adjustment is correct when no end play exists and a barely perceptible drag is felt when turning the steering wheel with thumb and forefinger. If adjustment is not as specified, unbolt the upper cover (29) from gear housing and vary the number of shims (28) until desired adjustment is obtained. Shims are available in thicknesses of 0.002, 0.003 and 0.010.

13B. LEVER SHAFT STUD MESH. With the cam shaft end play adjusted as in paragraph 13A, turn the steering gear to the mid or straight ahead posi-

tion and tighten adjusting screw (9— Fig. IH606A), located in (trunion) cover (23) on right side of housing, until a slight drag is felt when turning the steering gear through the mid or straight ahead position. Tighten the adjusting screw lock nut (8).

13C. GEAR SHAFT END PLAY. With the cam shaft end play and the lever shaft mesh adjusted as outlined in paragraphs 13A and 13B, turn the steering gear to the mid or straight ahead position and tighten adjusting screw (9—Fig. IH606A) to remove all end play from gear shaft (13) without increasing the amount of pull required to turn the steering gear through the mid or straight ahead position.

13D. GEAR TEETH MESH. With the preceeding adjustments completed and the gear unit in the mid or straight ahead position, grasp the steering (Pitman) arm on left side of tractor and hold this arm stationary while attempting to move the arm on right side, back and forth, to determine the amount of gear backlash. Note: The arm on the left side must be held stationary to avoid confusing

the backlash between the lever stud and the cam with the backlash between the gear teeth. If no backlash exists, the mesh position of the gear teeth can be considered satisfactory. If backlash does exist, the gear unit should be removed and overhauled as in paragraph 14B.

14. REMOVE AND REINSTALL To remove the steering gear unit first drain cooling system and remove hood, battery and starting motor. Disconnect the heat indicator sending unit, fuel lines, oil pressure gage line, wiring harness and controls from engine and engine accessories. Remove air cleaner and disconnect wires from head lights. Disconnect tail light wires and disconnect drag links from the steering (Pitman) arms. Unbolt steering gear housing and fuel tank from tractor and using a hoist, lift the fuel tank, instrument panel and steering gear housing assembly from tractor.

14A. Remove the steering wheel retaining nut and using a suitable puller, remove the steering wheel. Unbolt and remove the instrument panel assembly and fuel tank.

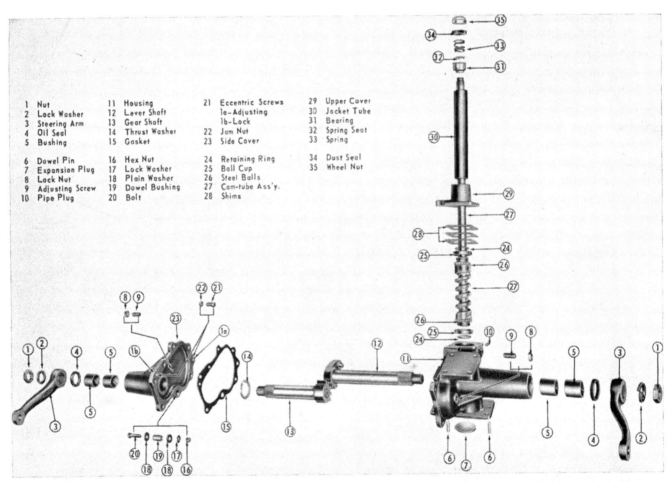

1	Nut	11	Housing	21	Eccentric Screws	29	Upper Cover
2	Lock Washer	12	Lever Shaft	1a-	Adjusting	30	Jacket Tube
3	Steering Arm	13	Gear Shaft	1b-	Lock	31	Bearing
4	Oil Seal	14	Thrust Washer	22	Jam Nut	32	Spring Seat
5	Bushing	15	Gasket	23	Side Cover	33	Spring
6	Dowel Pin	16	Hex Nut	24	Retaining Ring	34	Dust Seal
7	Expansion Plug	17	Lock Washer	25	Ball Cup	35	Wheel Nut
8	Lock Nut	18	Plain Washer	26	Steel Balls		
9	Adjusting Screw	19	Dowel Bushing	27	Cam-tube Ass'y.		
10	Pipe Plug	20	Bolt	28	Shims		

Fig. IH606A—Exploded view of the 300U, 350U and 350DU manual steering gear. Camshaft end play is controlled by shims (28). Lever shaft mesh position and gear shaft end play is controlled by screws (9). All bushings are available for service installation.

14B. **OVERHAUL.** To overhaul the steering gear, first remove the unit from tractor as outlined in paragraphs 14 and 14A. Remove the Pitman arm retaining nuts (1—Fig. 1H-606A) and using a suitable puller, remove the steering (Pitman) arms (3) from the lever shaft and gear shaft. Unbolt the side cover (23) from gear housing and remove the side cover and gear shaft (13). Withdraw lever shaft (12). Unbolt and remove the housing upper cover (29) and save shims (28) for reinstallation. Withdraw cam shaft (27) and remove ball cups (25) by removing their retaining snap rings (24). Thoroughly clean and examine all parts for damage or wear. The lever shaft and gear shaft should be renewed if the spur gear teeth are damaged or worn.

Inside diameter (new) of the lever shaft and gear shaft bushings is 1.374-1.375. Diameter of lever shaft and gear shaft at bearing surfaces is 1.3725-1.3735. Renew the shafts and/or bushings if running clearance is excessive.

New bushings should be pressed into position with a suitable piloted arbor until outer ends of bushings are flush with inner edge of chamfered surface in bores. Factory recommendations state that bushings should be burnished after installation to an inside diameter of 1.374-1.375.

When installing the cam shaft and jacket tube, vary the number of shims (28) to remove all camshaft end play and provide a barely perceptible drag when turning steering wheel with thumb and forefinger. Install the lever shaft, gear shaft and side cover and be sure thrust washer (14) is installed between gear shaft and side cover. Install the steering (Pitman) arms.

NOTE: In various periods of production, different length steering (Pitman) arms have been used. Also, there may be assembly chisel marks on the housing, arms and shafts; these marks, however, should not be used as the final index for correct assembly. When installing the steering (Pitman arms), turn the steering gear to the mid or straight ahead position and refer to Fig. IH606B where the installation dimensions for both the long and short arms are shown. If the assembly chisel marks do not align when arms are installed as per the dimensions in Fig. IH606B, grind-off the old marks and affix new ones. These new marks can then be used to locate the straight ahead position when the front wheel tread is subsequently changed.

14C. Now, with the steering gear in the mid or straight ahead position and the steering (Pitman) arms installed, check for backlash between the spur teeth on the lever shaft and gear shaft as follows: Grasp the steering arm (3) on the lever shaft (12) and hold this arm stationary while attempting to move the arm (3) on the gear shaft (13), back and forth, to determine the amount of gear backlash. Note: The arm on the lever shaft (12) must be held stationary to avoid confusing the backlash between the lever stud and the cam with the backlash between the spur teeth. If no backlash exists, the mesh position of the spur teeth can be considered satisfactory. If backlash does exists, and the lever shaft and gear shaft are known to be in new or satisfactory condition, proceed as follows: Remove the gear housing side cover, then remove the locating dowel from hole (X—Fig. IH606C) and blind hole (Y), but do not remove the pivot dowel shown in Fig. IH606D. Loosen the eccentric adjusting screw and locking screw (Fig. IH606C); then reinstall

the housing side cover and tighten the retaining screws enough to permit cover to move but not slip. Make sure also that the thrust adjusting screws (9) are loosened three or four turns.

With the steering gear in the mid-position, turn the eccentric adjusting screw clockwise to remove all backlash without causing any binding in the spur teeth when checked as before. Tighten the adjusting screw lock nut. Then lock the adjustment with the eccentric locking screw and tighten its jam nut. Tighten all other cover screws and bolts.

After the spur teeth mesh position is adjusted and the cover screws securely tightened, it will be necessary to hand ream the dowel bushing holes for oversize dowels which are available for service installation. The rear dowel seats in a blind hole and unless a suitable reamer is available, this dowel may be omitted. The upper front hole (X) should be reamed as shown in Fig. IH606D for a press fit of the oversize dowel.

Complete the overhaul by adjusting the lever stud mesh as in paragraph 13B and the gear shaft end play as outlined in paragraph 13C.

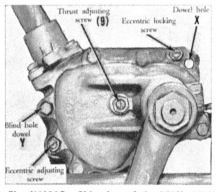

Fig. IH606C—Side view of the 300U, 350U and 350DU manual steering gear, showing the location of the eccentric screws used for adjusting the mesh between the spur teeth on the lever shaft and gear shaft.

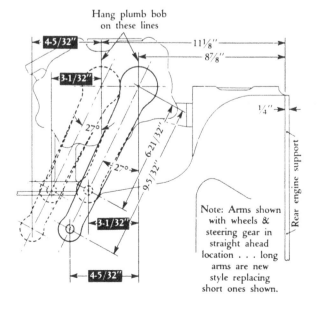

Fig. IH606B—Installation dimensions for the steering (Pitman) arms on the 300U, 350U and 350DU. Both the long and short arms are shown.

Note: Arms shown with wheels & steering gear in straight ahead location . . . long arms are new style replacing short ones shown.

Fig. IH606D—On 300U, 350U and 350DU models, it is necessary to ream the dowel holes for installation of oversize dowels after adjusting the backlash between the spur teeth on the lever shaft and gear shaft.

POWER STEERING SYSTEM

NOTE: The maintenance of absolute cleanliness of all parts is of utmost importance in the operation and servicing of the hydraulic power steering system. Of equal importance is the avoidance of nicks or burrs on any of the working parts.

LUBRICATION AND BLEEDING
All Models

15. The regular Hydra-Touch system fluid reservoir is the source of fluid supply to the power steering system. Only IH "Touch Control" or "Hydra-Touch and Touch Control" fluid should be used in the hydraulic system and the reservoir fluid level should be maintained at the "Full" mark on dip stick. Whenever the power steering oil lines have been disconnected, reconnect the oil lines, fill the reservoir and cycle the power steering system several times to bleed air from the system; then, refill the reservoir to "Full" mark on dip stick.

OPERATING PRESSURE, RELIEF VALVE, FLOW CONTROL VALVE
All Models With Hydra-Touch System

17. Working fluid for the hydraulic power steering system is supplied by the same pump which powers the Hydra-Touch system. Interposed between the pump and the Hydra-Touch system is a flow control valve mechanism which is shown schematically in Fig. IH606E. The small metering hole in the end of the flow valve piston passes between 2½ to 3 gallons per minute to the power steering system; but, since the pump supplies considerably more than three gpm, pressure builds up in front of the piston and moves the piston, against spring pressure, until the ports which supply oil to the Hydra-Touch system are uncovered. The power steering system, therefore, receives priority and the fluid requirements of the steering system are satisfied before any oil flows to the Hydra-Touch system. The auxiliary safety valve for the power steering system maintains a system operating pressure of 1200-1500 psi. The components of the flow control valve are shown exploded from the valve housing in Fig. IH606F.

17A. A pressure test of the power steering circuit will disclose whether the pump, safety valve or some other unit in the system is malfunctioning. To make such a test, proceed as follows: Connect a pressure test gage and shut-off valve in series with the line connecting the flow control valve to the steering valves as shown in Fig. IH 606G. Notice that the pressure gage is connected in the circuit between the shut-off valve and the flow control valve. Open the shut-off valve and run engine at low idle speed until oil is warmed. Advance the engine speed to the specified high idle rpm, close the shut-off valve, observe the pressure gage reading, then open the shut-off valve. If the gage reading is between 1200 and 1500 psi with the shut-off valve closed, the hydraulic pump and auxiliary safety valve are O. K. and any trouble is located elsewhere in the system.

If the gage reading is more than 1500 psi, the auxiliary safety valve may be stuck in the closed position. If the gage reading is less than 1200 psi, renew the auxiliary safety valve spring and recheck the pressure reading. If the gage reading is still less than 1200 psi, a faulty hydraulic pump is indicated.

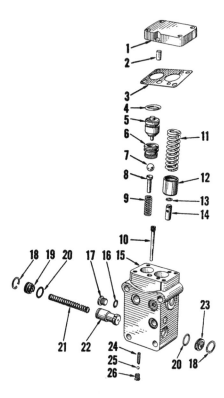

Fig. IH606F—Exploded view of the regulator, safety and flow control valve used on all models with Hydra-Touch system and power steering. The auxiliary safety valve (24, 25 and 26) protects only the power steering system.

Valve (22) is not available separately.

1. Cover
2. Dowel pin
3. Gasket
4. Seal ring
5. Regulator valve piston
6. Regulator valve seat
7. Steel ball
8. Ball rider
9. Ball rider spring
10. Safety valve orifice screen and plug
11. Safety valve spring
12. Spring retainer
13. Snap ring
14. Safety valve piston
15. Valve housing
16. Seal ring
17. Plug
18. Snap ring
19. Retainer
20. Seal ring
21. Flow control valve spring
22. Flow control valve
23. Retainer
24. Auxiliary safety valve spring
25. Steel ball
26. Plug

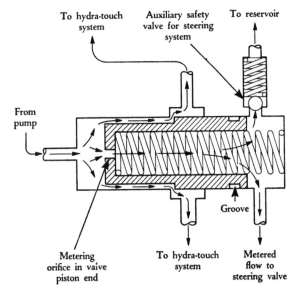

Fig. IH606E — Schematic illustration of the power steering flow control valve used on all models equipped with Hydra-Touch system. The flow control valve satisfies the 2½-3 gpm requirement of the power steering system before any oil flows to the Hydra-Touch system.

Series W400-W400D-W450-W450D
Without Hydra-Touch

17B. Working fluid for the hydraulic power steering system is supplied by a belt driven Eaton rotor type pump.

17C. A pressure test of the power steering circuit will disclose whether the pump, relief valve or some other unit in the system is malfunctioning. To make such a test, proceed as follows:

Connect a pressure test gage and shut-off valve in series with the pump pressure lines as shown in Fig. IH607. Notice that the pressure gage is connected in the circuit between the shut-off valve and the pump. Open the shut-off valve and run the engine

Fig. IH607B — Series W400, W400D, W450 and W450D (without Hydra-Touch) power steering relief valve (6A) and relief valve spring (7) can be removed from the flow control valve (6) after removing snap ring (5).

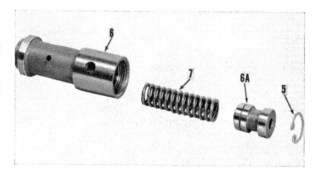

at low idle speed until the hydraulic fluid is warmed. Advance the engine speed to high idle rpm, close the shut-off valve and retain in the closed position only long enough to observe the gage reading. Note: Pump may be seriously damaged if valve is left in

the closed position for more than 3-4 seconds. If the gage reading is 800-850 psi, with the shut-off valve closed, the pump and relief valve are O. K. and any trouble is located in the control valve, power cylinder and/or connections.

If the gage pressure is more than 850 psi, the relief valve is probably stuck in the closed position. If the gage pressure is less than 800 psi, renew the relief valve spring (7—Fig. IH607A) and recheck the pressure reading. If the gage pressure is still too low, it will be necessary to overhaul the pump as outlined in paragraph 18A.

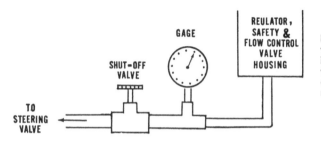

Fig. IH606G — Shut-off valve and pressure gage installation diagram for trouble shooting the power steering system on all models with Hydra-Touch system.

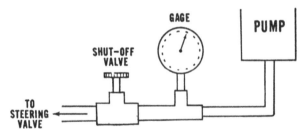

Fig. IH607 — Pressure gage and shut-off valve installation for trouble shooting the power steering system on series W400, W400D, W450 and W450D not equipped with Hydra-Touch.

PUMP
All Models With Hydra-Touch

18. The regular Hydra-Touch pump supplies fluid to both the power steering system and the Hydra-Touch system. Refer to paragraph 223 for R&R and/or resealing of the pump.

Series W400-W400D-W450-W450D
Without Hydra-Touch

18A. OVERHAUL. With the pump removed from tractor, refer to Fig. IH607A and proceed as follows: Thoroughly clean exterior of pump in a suitable solvent and remove the drive pulley and key. Then unbolt and separate cover from pump body (Tap gently if necessary). Mark the rotors so they can be reinstalled in the same relative position and remove the rotors and drive key from pump shaft. Remove the bearing retaining snap ring (20) and press the pump shaft and bearing from pump body. Remove valve cap (1) and withdraw the flow control valve spring (3), orifice plate (4) and flow control and relief valve assembly (5, 6, 6A and 7). Remove snap ring (5) and extract relief valve (6A) and spring (7). Refer to Fig. IH607B.

Wash all parts EXCEPT the rotor shaft bearing in a suitable solvent and inspect. If bushing in pump cover is worn, renew the cover because the bushing is not available separately. Renew any other questionable parts and check the clearances as follows:

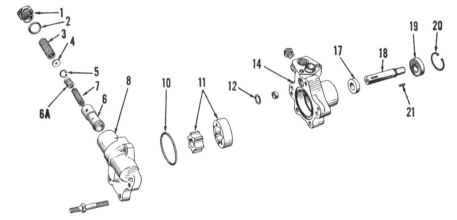

Fig. IH607A—Exploded view of the power steering pump used on the W400, W400D, W450 and W450D series without Hydra-Touch. Rotors (11) are available in matched sets only. Valves (6 or 6A) are not available separately.

1. Valve cap	6. Flow control valve	14. Pump body
2. Seal ring	6A. Relief valve	17. Oil seal
3. Flow control valve spring	7. Relief valve spring	18. Shaft
4. Orifice plate	8. Pump cover	19. Ball bearing
5. Snap ring	10. "O" ring gasket	20. Snap ring
	11. Rotors	21. Dowel pin
	12. "O" ring	

Insert rotor shaft and bearing into pump housing until bearing is in position. Install rotor key and rotors and check the clearance between the rotor teeth as shown in Fig. IH607C. Note: Rotors must be installed in housing when making this check. If clearance exceeds 0.006, renew the set of rotors. Check the clearance between the driven rotor and the body bushing as shown in Fig. IH607D. Recommended clearance is 0.001-0.0025. Check the end clearance of rotors as shown in Fig. IH607E. If the end clearance exceeds 0.002, renew the pump body and/or rotors set.

Thoroughly dry the relief valve (6A—Fig. IH607B) and bore of flow control valve and make sure the relief valve slides freely in its bore. Any burrs restricting the free movement of the valve can be removed with crocus cloth. Check freedom of movement of flow control valve (6) in the pump cover bore in the same manner.

Before assembly, coat all parts in power steering fluid. If new rotor shaft seal is installed, coat lip of same with Lubriplate or equivalent and install seal with lip of same toward rotors. Renew all other "O" ring seals and reassemble.

STEERING VALVES
Series 300U-350U-350DU

19. R&R AND OVERHAUL. To remove the power steering valves, it is first necessary to remove the complete steering gear unit from tractor. The procedure for removing the power steering gear unit is the same as the procedure for removing the manual gear unit, outlined in paragraph 14, except it is necessary to perform the additional work of draining the hydraulic system and disconnecting the power steering oil lines.

With the gear unit removed from tractor, unbolt and remove the jacket tube and steering valve upper cover assembly. Refer to Fig. IH608. Unbolt and remove the bearing adjusting nut and lift out the upper thrust bearing.

Fig. IH607D—Checking clearance between driven rotor and body bushing on series W400, W400D, W450 and W450D (without Hydra-Touch) power steering pumps.

Disconnect the oil lines from valve body and withdraw the valve assembly and lower thrust bearing. Be careful when removing the valve assembly and do not drop or nick any of the component parts.

Carefully slide the spool from the valve body and remove the six active plungers and the three centering springs as shown in Fig. IH608A. The two inactive plungers can be pressed from the valve body at this time. Thoroughly clean all parts in a suitable solvent and be sure all passages and bleed holes are open and clean. Note: On some early models, there was a bleed hole drilled through one of the inactive plungers. On later models, this bleed hole is drilled through the valve body and both of the inactive plungers are solid. CAUTION: Never substitute a solid inactive plunger for one that is drilled.

The valve spool and body are mated parts and must be renewed as an assembly if damaged. Each of the centering springs should have a free length of 0.703 and should test 11-15 pounds when compressed to a height of 0.638 inches. The plungers and springs are available separately.

When reassembling the control valves, be sure to renew all "O" rings and seals and proceed as follows: Install the lower race of the lower thrust bearing in the adapter casting with ball groove up. Pack the ball bearing with wheel bearing grease and install bearing. Then, install upper race of lower thrust bearing with ball groove down. Press the two long inactive plungers into the valve body with one plunger hole between them as shown in Fig. IH608B. Refer to Fig. IH608A and install the six active plungers and the three centering springs in the remaining three plunger holes in valve body. Install the control spool in the valve body so that the identification groove in I. D. of

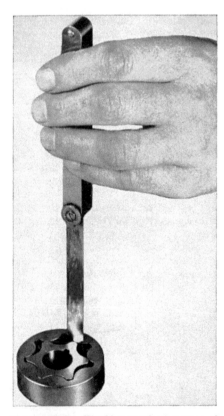

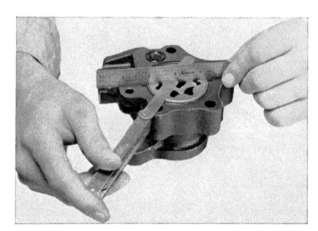

Fig. IH607E — Checking rotors end clearance on series W400, W400D, W450 and W450D (without Hydra-Touch) power steering pumps.

Fig. IH607C—Checking tooth clearance of rotors used in Series W400, W400D, W450 and W450D (without Hydra-Touch) power steering pump. Rotors must be installed in pump body when making this check.

spool is toward the same side of body as the port identification symbols "PR" and "RT". Refer to Fig. IH608C. Then install the assembled valve body with the symbols "PR" and "RT" up toward steering wheel. Install the upper thrust bearing with larger diameter race toward valve body and be sure to pack the bearing with wheel bearing grease. Install the tongued lockwasher and nut, tighten the nut just enough to hold the assembly from slipping and center the upper thrust bearing lower race as follows:

Using a pair of dividers or six-inch scale, measure the distance from outer edge of lower race to outside edge of each of the five plunger holes. Shift the bearing race until these five meas-

Fig. IH608A—Power steering control valve for series 300U, 350U and 350DU, with the control spool, active plungers and centering springs removed. The inactive plungers can be pressed from the valve body.

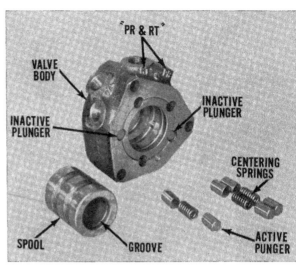

urements are identical. NOTE: The bearing race must be perfectly centered to prevent subsequent binding of the valve spool. Before tightening the adjusting nut, make certain that the lever shaft stud mesh, gear shaft end play and rack mesh are properly adjusted as in paragraph 25A, 25B and 25D.

Fig. IH608B—Installing the inactive plungers in series 300U, 350U and 350DU power steering valve body. There must be one plunger hole (for active plungers) between the two inactive plungers as shown.

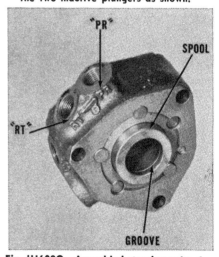

Fig. IH608C—Assembled steering valve for series 300U, 350U and 350DU. The spool should be installed so that groove in I.D. of same is toward same side of valve body as the cast in "PR & RT" markings.

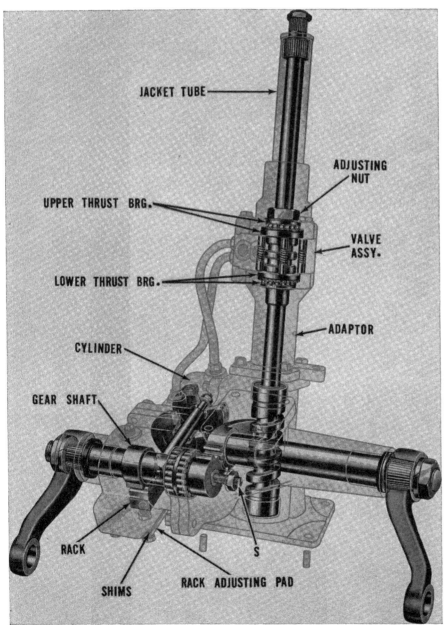

Fig. IH608—Phantom view of series 300U, 350U and 350DU power steering gear, cylinder and control valve unit. The system working fluid is supplied by the regular Hydra-Touch system.

Install the steering wheel loosely on the shaft serrations and turn the steering wheel to the left, off the mid or straight ahead position, to lift the steering shaft to the upper extreme. Hold the steering wheel in this left turn position and tighten the valve adjusting nut to a torque of 10-12 Ft.-Lbs. If a suitable torque wrench is not available, a pair of ten inch multi-slip joint pliers can be used. Turn the steering wheel to both extreme positions several times and recheck the nut adjustment with the shaft raised or in the left turn position. Back-off the nut 1/12 turn (½ of a hex face) and bend two of the washer prongs against flats of nut.

Install the jacket tube assembly, mount a dial indicator and check the up and down movement of the steering (cam) shaft. When turning from the mid or straight ahead position to the extreme left, the shaft should move upward 0.050-0.055. When turning from the mid or straight ahead position to the extreme right, the shaft should move downward 0.050-0.055. In other words, when turning the steering wheel from one extreme position to the other, the shaft should have a total movement of 0.100-0.110. If the shaft movement is not as specified, it will be necessary to recheck the lever shaft stud mesh position and the valve nut adjustment.

Complete the assembly by installing the oil lines, instrument panel and steering wheel. Tighten the steering wheel retaining nut to a torque of 30-40 Ft.-Lbs.

Series W400-W400D-W450-W450D

20. **ADJUSTMENT.** To adjust the cylinder and valve unit, loosen the clamp on the steering drag link, then unstake and loosen the knurled ad-

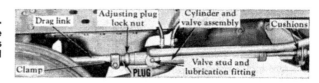

Fig. IH609A—Power steering cylinder and valve unit installation on series W400, W400D, W450 and W450D.

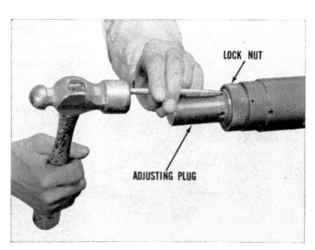

Fig. IH609B—Series W-400, W400D, W450 and W450D, power steering valve adjusting plug can be unscrewed from body after unstaking and removing the knurled lock nut.

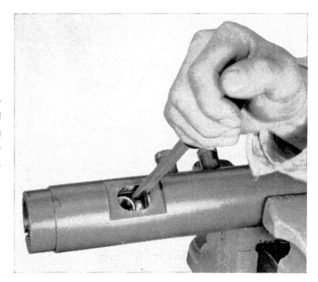

Fig. IH609C—Series W-400, W400D, W450 and W450D, power steering valve can be pushed out of housing, using a large screw driver as shown.

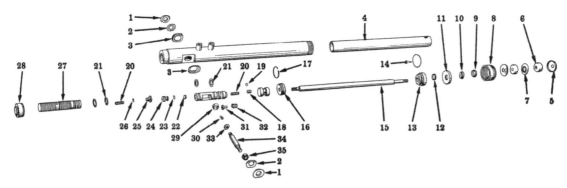

Fig. IH609—Exploded view of series W400, W400D, W450 and W450D power steering cylinder and valve unit. If the valve spool, cylinder inner tube and/or housing are damaged, the complete unit should be renewed.

1. Stud cushion	8. Cylinder tube cap	16. Piston	23. Plug seal ring	28. Knurled lock nut
2. Thrust washer	9. Wiper ring retainer	17. Piston ring	24. Spool adjusting plug	29. Valve stud ball
3. Cushion retainer	10. Wiper ring	18. Inner reaction piston	25. Outer reaction piston	30. Lubrication fitting
4. Cylinder inner tube	11. T-ring retainer	19. "O" ring	26. "O" ring	31. Inner ball seat
5. Cushion retainer	12. T-ring and washer	20. Reaction piston springs	27. Valve adjusting plug	32. Spool spring
6. Frame bracket cushion	13. Cylinder rod guide	21. Valve seal rings		33. Elastic stop nut
7. Cushion locator	14. Guide seal ring	22. Outer ball seat		34. Valve stud
	15. Piston rod			35. Slotted nut

justing plug lock nut. Refer to Fig. IH609A. Insert a ¼-inch rod in hole in the adjusting plug and turn the plug not more than 1/6-turn either way as required; then road test tractor to determine if additional adjustment is needed.

NOTE: Turn the adjusting plug out to correct the following:

    a. Right turn is too hard
    b. Left turn is too easy
    c. Poor recovery after right turns.
    d. Tractor wanders to the left

Turn the adjusting plug in to correct the following:

    a. Left turn is too hard
    b. Right turn is too easy
    c. Poor recovery after right turns
    d. Tractor wanders to the right

20A. OVERHAUL. With the cylinder and valve unit removed from tractor, mark the relative position of the drag link with respect to the valve adjusting plug so they can be reassembled in the same position, loosen the clamp and unscrew drag link from the adjusting plug. Remove the through stud (34 — Fig. IH609) and withdraw the valve control arm. Using a small chisel as shown in Fig. IH609B, unstake and remove the knurled lock nut; then, mark the depth to which the adjusting plug is screwed into the housing and unscrew the plug from the housing. Using a large screw driver as shown in Fig. IH609C, push the valve assembly out of housing, but be careful not to scratch or otherwise damage any of the polished surfaces. Pull out the reactor pistons (18 & 25—Fig. IH609) and remove the springs (20). Unscrew and remove plug (24) and disassemble the remaining parts. Thoroughly clean all parts in a suitable solvent and renew any which are damaged or show wear. If the valve spool or spool bore is damaged, the complete cylinder and valve unit should be renewed.

When reassembling, be sure to renew all "O" ring seals and lubricate all parts in power steering fluid before installation. Install spool spring (32), inner ball seat (31), ball (29), outer ball seat (22) and adjusting plug (24). Turn the adjusting plug (24) completely in, then back the plug off ¼-turn and stake. Install the reactor springs (20) and reactor pistons (18 & 25). Install the assembled spool and screw the adjusting plug in to its original position. Install the knurled lock nut but do not tighten. Screw the drag link on the adjusting plug to its

original position. Install the cylinder and valve unit on the tractor and adjust the unit as outlined in paragraph 20. Then tighten the drag link clamp and tighten and stake the knurled lock nut.

STEERING CYLINDER
Series 300U-350U-350DU

21. R&R AND OVERHAUL. To remove the power steering cylinder, first drain the hydraulic system and remove the battery, generator regulator and the hydraulic manifold connecting pump to reservoir. Unbolt and remove the rack adjusting pad and shims from bottom of gear housing and save the shims for reinstallation. Refer to Fig. IH608. Disconnect oil lines from cylinder, then unbolt and remove cylinder and rack assembly from gear housing.

Using a plastic or lead hammer, bump the cylinder off the cylinder end plate and piston. Remove the nut retaining piston to piston rod, remove piston and withdraw the piston rod from the cylinder end plate. The procedure for further disassembly is evident. Thoroughly clean all parts in a suitable solvent and renew any damaged or worn parts. Be sure to examine rubbing surfaces of the gear rack and adjusting pad for excessive wear. Be sure to renew all rings, seals and gaskets.

When reassembling, lubricate all parts in power steering fluid and proceed as follows: Slide the piston rod through the cylinder end plate and install piston so that cupped end will be toward the retaining nut. Install the self-locking nut and tighten same securely as shown in Fig. IH610. Install piston, end plate and rod assembly into cylinder. Check the lever stud

mesh postion as in paragraph 25A and the gear shaft end play as in paragraph 25B, then install the assembled cylinder so that the first tooth of the gear sector meshes with the first space in the power rack.

Install the rack adjusting pad and vary the number of shims between the pad and the housing to provide a slight drag between the rack teeth and gear shaft teeth when the pad retaining screws are securely tightened; then add one 0.003 shim, coat shims with gasket sealer and reinstall the adjusting pad.

Note: Some early production tractors were equipped with only one ¼-inch thick pad retaining plate. When such tractors are encountered, install one more ¼-inch thick plate and use ¼-inch longer retaining screws to take care of the added thickness of the two plates. On some late models, the mesh adjusting shims are located between adjusting pad and pad plate.

Complete the assembly by installing the oil lines, battery and generator regulator.

Series W400-W400D-
W450-W450D

22. OVERHAUL. With the cylinder and valve unit removed from tractor, unscrew the cylinder tube cap (8— Fig. IH609) and pull piston and rod assembly from the cylinder tube. Thoroughly clean all parts and examine them for damage or wear.

NOTE: The cylinder is fitted with an inner tube (4) which is pressed into the rear valve retainer. If this inner tube is damaged, it is recommended that the complete cylinder and valve unit be renewed.

Piston ring should not bind in the piston grooves and rings should have an end gap of 0.001-0.006, when checked as shown in Fig. IH611.

Fig. IH610—Series 300U, 350U and 350DU, power steering cylinder. Piston should be installed with cupped side toward the retaining nut.

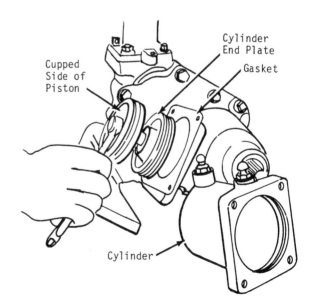

Cylinder End Plate

Gasket

Cupped Side of Piston

Cylinder

When reassembling, lubricate all parts in power steering fluid and renew all seals. Using a suitable ring compressor, install the piston and rod assembly as shown in Fig. IH611A. Tighten the cylinder end cap securely.

GEAR MOTOR & STEERING VALVE UNIT
Series 300-350-350D-400-400D-450-450D

23. **REMOVE AND REINSTALL.** To remove the power steering unit, first remove the hood and grille, loosen the steering shaft coupling located in front of the steering unit and slide the coupling forward on the worm shaft. Refer to Fig. IH612. Remove the steering worm shaft front bearing retainer and turn the worm shaft forward and out of the steering gear housing. Remove the air cleaner and remove the strap iron support braces. Turn the power unit until the oil line connections are up and disconnect the oil lines. Loosen the clamp bolts in the steering shaft U-joint, slide the power unit forward and remove the U-joint Woodruff key. Then, slide the unit forward and remove same from tractor.

23A. **OVERHAUL.** With the power unit removed from tractor, thoroughly clean same in a suitable solvent to remove any accumulation of dirt or other foreign material, refer to Fig. IH612A and proceed as follows: Affix index marks to the valve body (25), bearing plate (28) and end cup (27) so they can be reassembled in the same position, remove cap screws (26) and lift off the end cup (27), but be careful not to remove the idler gear and shaft (22). Also be careful not to damage or lose the gasket (21). Use chalk or crayon and mark the meshing teeth of the idler gear and power gear so the same teeth can be remeshed during assembly; then remove the idler gear and shaft. Remove any paint from the exposed end of the steering shaft, then withdraw the steering shaft, bearing plate and valve assembly from the valve body as shown in Fig. IH612B. Be careful not to damage or lose the body gasket (or gaskets) (23—Fig. IH612A).

Note: If the unit is being disassembled for the purposes of renewing seals only, the removed bearing plate, steering shaft and valve assembly need not and should not be disassembled. If, however, the unit is to be disassembled, proceed as follows:

Using a suitable punch, drift out the roll pin (47—Fig. IH612A) retaining cam sleeve (43) to power gear and shaft (42). then remove the power

gear and shaft and bearing plate (28). Clamp the cam sleeve (43) in a soft jawed vise, turn steering shaft until center hole in cam sleeve is in register with center pin (46) and drift out the pin (46). Pull steering shaft and valve assembly out of cam sleeve as far as possible and, using a small punch, drift out the pin (31—Fig. IH612C). Then, push the steering shaft into the cam sleeve and turn the steering shaft until hole (from which pin (31) was removed) in shaft is aligned with pin (45) and drift pin (45) from the cam sleeve. Withdraw steering shaft from cam sleeve, catching the steel balls (29—Fig. IH612A) and spring (30) as they fly out.

Thoroughly clean all parts and renew any which are damaged or worn.

End cup casting (27), bearing plate (28) and gears (22 and 42) are available as matched units only. If any one of the parts is damaged, it will be necessary to renew all four. Similarly, the valve body casting (25), valve spool (39) and spool sleeve (41) are available as matched units only. All other parts are available separately if desired and the procedure for renewing them is evident. Lip of wiper seal (33) should be installed away from valve body. Make sure that spring (30) and steel balls (29) are in good condition.

Examine the valve body casting near the bearing plate location for stamped letters "S" and "F" as shown in Fig. IH612D. The "S" indicates that the

Fig. IH611—Checking the piston ring end gap on series W400, W400D, W-450 and W450D, power steering cylinder. Recommended end gap is 0.001-0.006.

Fig. IH611A — Using a ring compressor to install series W400, W400D, W-450 and W450D, power steering piston. All parts should be lubricated in power steering fluid prior to assembly.

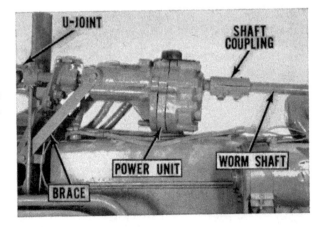

Fig. IH612 — Installation view of the Behlen power steering unit.

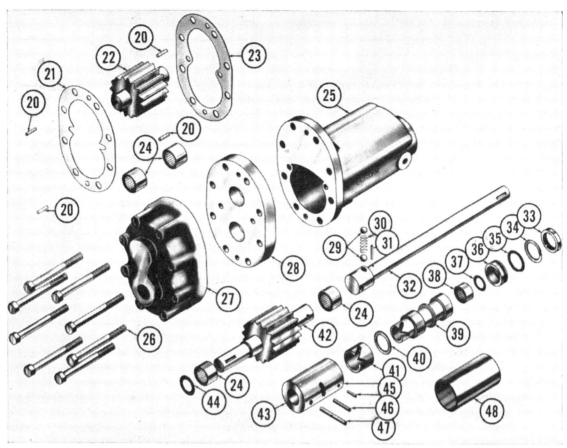

Fig. IH612A—Exploded view of the Behlen power steering unit. Hydraulic power for the steering system is supplied by the regular Hydra-Touch system pump.

20. Dowel pin
21. End cup gasket (0.005, 0.006, 0.007 and 0.008)
22. Idler gear and shaft
23. Body gasket (0.015)
24. Needle bearings
25. Valve body casting
26. Assembly screws
27. End cup
28. Bearing plate
29. Steel ball
30. Centering spring
31. Roll pin
32. Steering shaft
33. Dust wiper
34. Snap ring
35. "O" ring
36. Spacer ring
37. "O" ring
38. Needle bearing
39. Valve spool
40. Spring washer
41. Spool sleeve
42. Power gear and shaft
43. Cam sleeve
44. "O" ring
45. Roll pin
46. Roll pin
47. Roll pin
48. Retainer sleeve

Fig. IH612C—When assembling the valve shift mechanism on Behlen power steering units, make certain the parts are in the positions shown and refer to text.

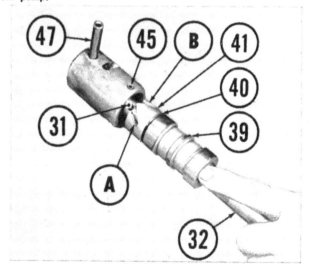

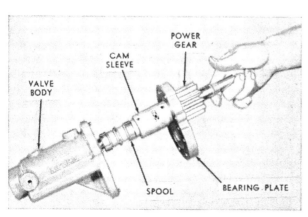

Fig. IH612B — Removing bearing plate and valve assembly from the Behlen power steering unit valve body.

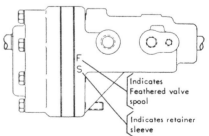

Fig. IH612D—"F" mark on power steering valve housing indicates that spool is feathered. "S" mark indicates that roll pin retainer sleeve has been installed. Refer to text.

power steering unit is equipped with roll pin retainer sleeve (48—Fig. IH-612A). If no "S" mark is found, obtain and install sleeve (48) and affix the "S" mark to the casting. The "F" indicates that the valve spool lands have been feathered to provide smoother, quieter operation. If the spool lands are not feathered, use a drill press and small round stone as shown in Fig. IH612E and feather the spool according to the dimensions shown in Fig. IH612F. Then affix the "F" mark to the casting. Note: The flats must not be ground deeper than 3/64-inch when measured axially as shown, or spool will be unsuitable for further use. Radial depth of flats is immaterial and will vary according to the diameter of the grinding stone used.

When reassembling, install centering spring (30—Fig. IH612A) and steel balls (29) in steering shaft, compress spring (30) and push steering shaft into cam sleeve. Turn the steering shaft either way from center position; at which time, spring (30) should have sufficient tension to snap the steering shaft back to center. Bump pin (45—Fig. IH612C) into cam sleeve until pin is flush with outside diameter of cam sleeve. Then pull the steering shaft outward from cam sleeve and bump pin (31) into the steering shaft just far enough to clear the inside diameter of the cam sleeve. Install spool sleeve (41), spring washer (40) and valve spool (39) on the steering shaft. Position the parts exactly as shown in Fig. IH612C. Notice that pin (31) is on left side of pin (45). CAUTION: Steering unit will not work if pin (31) is on right side of pin (45).

Now, push inward on steering shaft and at the same time, turn shaft clockwise until pin (31) engages slots (A) in spool valve (39) and spool sleeve (41). Then, while still pushing inward, turn steering shaft counterclockwise until pin (45) engages slots

(B). Now, push the steering shaft and valve shift mechanism completely into the cam sleeve, align pin holes and install center pin (46—Fig. IH-612A). Install retainer sleeve (48). Install bearing plate (28) and power gear (42); then, install roll pin (47) so that slit in pin is toward power gear.

If the original gasket (or gaskets) (23) are in good condition, install it (or them) on valve body, lubricate the assembled valve unit in clean power steering fluid and install the unit in the valve body as shown in Fig. IH-612B.

Note: The number of gaskets (23) determines whether the valve spool is centered in the valve body. To check the valve spool position, remove the fitting from the pressure inlet port in the valve body, and while holding the

bearing plate and gaskets tightly against the valve body, sight through the inlet port and make certain that the middle spool land is centered in the port within 0.015 either way. If spool is not centered, it will be necessary to vary the number of 0.015 thick gaskets (23).

Install the idler gear and shaft (22) and be sure to mesh the previously marked teeth on the power and idler gears.

Note: If new gears, bearing plate and end cup are being installed, the meshing gear teeth are marked with blue on the bearing plate side.

The thickness of gasket (21) controls the end play of the motor gears. If original gasket (21) is in good condition and if the original gears, bearing plate and end cup are used, install the original gasket and end cup.

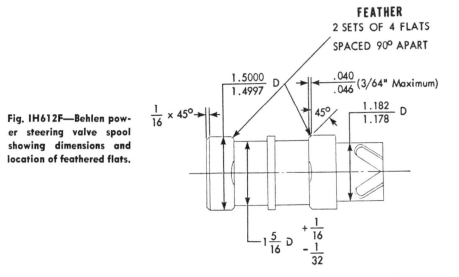

Fig. IH612F—Behlen power steering valve spool showing dimensions and location of feathered flats.

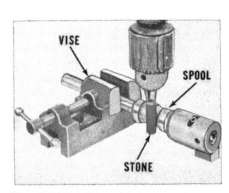

Fig. IH612E—Using drill press, vise and round cylindrical stone to feather the power steering valve spool on Behlen units.

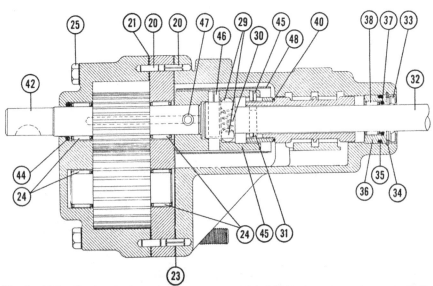

Fig. IH612G—Sectional view of properly assembled Behlen power steering unit. Refer to Fig. IH612A for legend.

If necessary to install a new gasket make certain that the new gasket is exactly the same thickness as the original one when checked with a micrometer. If new gears, bearing plate and end cup are being installed, be sure to use the gasket which is shipped with the new units. Gaskets are available in thicknesses of 0.005, 0.006, 0.007 and 0.008.

Tighten cap screws (26) to a torque of 50 ft.-lbs.

GEAR UNIT

Series 300-350-350D-400-400D-W400-W400D-450-450D-W450-W450D

24. The steering gear unit used on models with power steering is the same as the unit used on models without power steering. Refer to paragraphs 7 and 8 for series 300, 350, 350D, 400, 400D, 450 and 450D and paragraphs 9, 10, 11 and 12 for series W400, W400D, W450 and W450D.

Series 300U-350U-350DU

25. **ADJUSTMENT.** The steering gear unit is provided with five adjustments: The mesh position between the lever shaft stud and cam shaft, the end play of the gear shaft and the power rack mesh position can all be adjusted without removing the steering gear unit from tractor. The mesh position between the lever shaft and gear shaft gear teeth is adjustable; however, this adjustment requires reaming and installation of oversize dowels and should not be done except during overhaul and when a new side cover, gear housing, lever shaft and/or gear shaft are installed. As shown in Fig. IH613, the cam (worm) shaft is mounted in bushings and requires no adjustment. The steering valve thrust bearings are adjusted by tightening the bearing adjusting nut to a torque of 10 - 12 Ft.-Lbs., but this should be done only when the valve unit is being serviced.

Before attempting to make any adjustments, first make certain that the gear housing is properly filled with lubricant, then disconnect the drag links from the steering (Pitman) arms.

25A. LEVER SHAFT STUD MESH. With the steering gear in the mid or straight ahead position, loosen the lock nut and back-off the gear shaft end play adjusting screw (S—Fig. IH-613) three or four turns. Then loosen the lock nut and tighten the lever shaft adjusting screw, located in (trunion) cover on right side of housing, until a slight drag is felt when turning the steering gear through the mid

or straight ahead position. Tighten the lever shaft adjusting screw lock nut.

25B. GEAR SHAFT END PLAY. With the lever shaft stud mesh position adjusted as outlined in paragraph 25A, turn the steering gear to the mid or straight ahead position and tighten adjusting screw (S—Fig. IH-613) to remove all end play from gear shaft without increasing the amount of pull required to turn the steering gear through the mid or straight ahead position.

25C. GEAR TEETH MESH. Remove the rack adjusting pad (Fig. IH613) to avoid confusing the rack tooth backlash with the tooth backlash between

the lever shaft and gear shaft. Grasp the steering (Pitman) arm on left side of tractor and hold this arm stationary while attempting to move the arm on right side, back and forth, to determine the amount of gear backlash. Note: The arm on the left side must be held stationary to avoid confusing the backlash between the lever stud and cam with the backlash between the gear teeth. If no backlash exists, the mesh position of the gear teeth can be considered satisfactory. If backlash does exist, the gear unit should be removed and overhauled as in paragraph 25G.

Refer to the following paragraph before installing the rack adjusting pad.

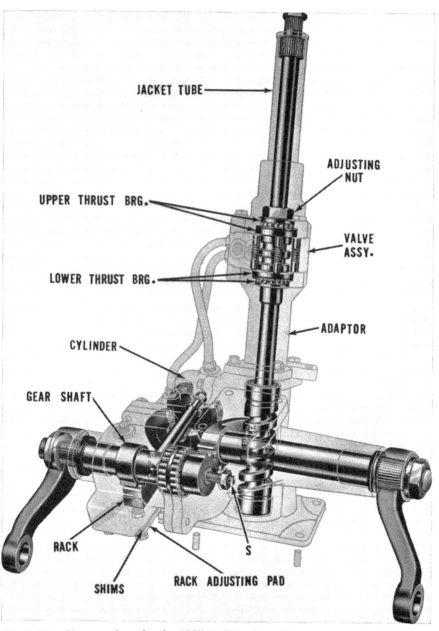

Fig. IH613—Phantom view of series 300U, 350U and 350DU power steering gear, cylinder and control valve unit. The system working fluid is supplied by the regular "Hydra-Touch" system.

25D. POWER RACK MESH. Refer to Fig. IH613. With the rack adjusting pad and shims removed from the gear unit, remove enough shims from between pad and housing to provide a slight drag between the rack teeth and gear shaft teeth when the pad retaining screws are securely tightened; then, remove pad and add one 0.003 shim, coat shims with gasket sealer and reinstall the adjusting pad.

Note: Some early production tractors were equipped with only one ¼-inch thick pad retaining plate. When such tractors are encountered, install one more ¼-inch thick plate and use ¼-inch longer retaining screws to take care of the added thickness of the two plates. On some late models, the mesh adjusting shims are located between adjusting pad and pad plate.

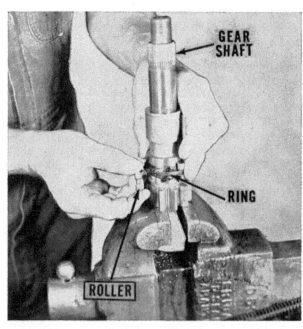

Fig. IH613C—Using heavy grease to assemble rollers and ring to power steering unit gear shaft on series 300U, 350U and 350DU.

25E. **REMOVE AND REINSTALL.** To remove the steering gear unit first drain cooling system and remove hood, battery and starting motor. Disconnect the heat indicator sending unit, fuel lines, oil pressure gage line, wiring harness and controls from engine and engine accessories. Remove air cleaner and disconnect wires from head lights.

Disconnect tail light wires and disconnect drag links from the steering (Pitman) arms. Disconnect the oil lines connecting the hydraulic oil reservoir to the steering valves. Unbolt steering gear housing and fuel tank, from tractor and using a hoist, lift the fuel tank, instrument panel and steering gear housing assembly from tractor.

25F. Remove the steering wheel retaining nut and using a suitable puller, remove the steering wheel. Unbolt and remove the instrument panel asesmbly and fuel tank.

25G. **OVERHAUL.** To overhaul the steering gear, first remove the unit from tractor as outlined in paragraphs 25E and 25F. Remove the steering (Pitman) arm retaining nuts and using a suitable puller, remove the Pitman arms from the steering lever shaft and gear shaft. Refer to paragraph 19 for removal, reinstallation and overhaul of the steering valves and to paragraph 21 for R&R and overhaul of the steering cylinder.

With the valves and cylinder removed, unbolt the side cover from gear housing and remove the side cover and gear shaft .Withdraw lever shaft and cam shaft. Thoroughly clean and examine all parts for damage or wear. The lever shaft and gear shaft should be renewed if the spur gear teeth are damaged or worn. Inspect also the roller bearing at upper end of the jacket tube and the gear shaft rollers and race. If any part of the roller bearing (outer race or rollers) is damaged, it will be necessary to renew the complete bearing, as component parts are not catalogued.

Inside diameter (new) of the lever shaft and gear shaft bushings is 1.374-1.375. Diameter of lever shaft and gear shaft at bearing surfaces is 1.3725-1.3735. Renew the shafts and/or bushings if running clearance is excessive. New bushings should be pressed into position with a suitable piloted arbor until outer ends of bushings is flush with inner edge of chamfered surface

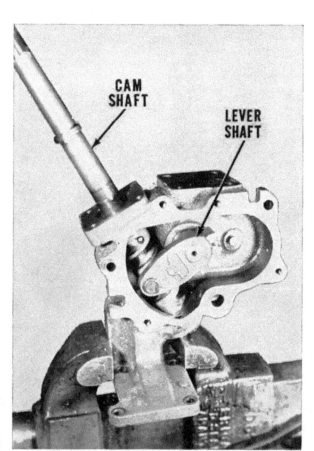

Fig. IH613B — Installing camshaft and lever shaft on series 300U, 350U and 350DU, power steering unit.

in bores. Factory recommendations state that bushings should be burnished after installation to an inside diameter of 1.374-1.375.

Inside diameter (new) of the cam shaft bushings located in gear housing is 1.6235-1.6250. Diameter of cam shaft at bushing surfaces is 1.620-1.621. Renew the cam shaft and/or bushings if running clearance is excessive. Factory recommendations state that bushings should be burnished after installation to an inside diameter of 1.6235-1.6250.

Install the lever shaft and cam shaft as shown in Fig. IH613B. Using heavy grease, assemble the rollers and retainer ring to gear shaft as shown in Fig. IH613C. Then install gear shaft in side cover as shown in Fig. IH613D. Install side cover and the steering (Pitman) arms.

NOTE: In various periods of production, different length steering (Pitman) arms have been used. Also, there may be assembly chisel marks on the housing, arms and shafts; these marks, however, should not be used as the final index for correct assembly. When installing the steering (Pitman arms),

turn the steering gear to the mid or straight ahead position and refer to Fig. IH606B where the installation dimensions for both the long and short arms are shown. If the assembly chisel marks do not align when arms are installed as per the dimensions in Fig. IH606B, grind-off old marks and affix new ones. These marks can then be used to locate the straight ahead position when the front wheel tread is subsequently changed.

Now, with the steering gear in the mid or straight ahead position and the steering (Pitman) arms installed, check for backlash between the spur teeth on the lever shaft and gear shaft in a similar manner to that described for the manual steering gear unit in paragraph 14C.

Complete the gear unit overhaul by adjusting the lever stud mesh as in paragraph 25A and the gear shaft end play as outlined in paragraph 25B. Then install the power cylinder and rack assembly and adjust the rack mesh position as in paragraph 25D. Refer to paragraph 19 when installing the steering valves and adjusting the bearing nut.

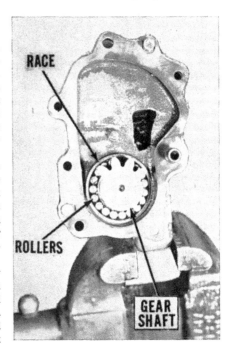

Fig. IH613D—Power steering gear shaft properly installed in housing side cover on series 300U, 350U and 350DU.

ENGINE AND COMPONENTS

R&R ENGINE WITH CLUTCH
Series 300-350-400-400D-450-450D

28. To remove the engine and clutch as an assembly, first remove hood; then on models with manual steering, disconnect the steering shaft universal joint and remove the U-joint Woodruff key. On models with power steering, loosen the steering shaft coupling which is located in front of the power unit. Disconnect the shutter control rod. Disconnect the radiator upper support bracket or rod, drain cooling system and disconnect the radiator hoses. If engine is to be disassembled, drain oil pan. Remove the dust shield from under front end of frame rails and on models with an adjustable type front axle, disconnect the radius rod at its rear ball joint. Support tractor under clutch housing, unbolt the upper bolster from the frame rails and move the bolster, radiator and front wheels assembly away from tractor.

Disconnect the heat indicator sending unit, fuel lines, oil pressure gage line, wiring harness and controls from engine and engine accessories. Remove the oil filter and on Diesel models, remove the starting control rod, fuel supply and return lines and carburetor

air cleaner pipe. Disconnect the hydraulic lines from the hydraulic pump.

Attach hoist to engine in a suitable manner and remove the frame side rails. Unbolt engine from clutch housing and move engine and clutch assembly forward and away from tractor.

Series W400-W400D-W450-W450D

29. To remove the engine and clutch as an assembly, first remove the hood and the radiator upper support rod. Disconnect the shutter control rod. Drain cooling system, disconnect radiator hoses and if engine is to be disassembled, drain oil pan. Disconnect the steering drag link from steering arm and the radius rod at its rear ball joint. Disconnect the head light wires. Support tractor under clutch housing, unbolt the upper bolster from the frame rails and move the bolster, radiator and front wheels assembly away from tractor.

Disconnect the heat indicator sending unit, fuel lines, oil pressure gage line, wiring harness and controls from engine and engine accessories. Remove the oil filter and on Diesel models, remove the starting control rod, fuel supply and return lines and carbure-

tor air cleaner pipe. Disconnect the hydraulic lines from the Hydra-Touch pump. On models equipped with a separate power steering pump, unbolt the pump from engine and without disconnecting hoses from pump, lay the unit rearward and out of way.

Attach hoist to engine in a suitable manner and remove the frame side rails. Unbolt engine from clutch housing and move engine and clutch assembly forward and away from tractor.

Series 300U-350U-350DU

30. To remove the engine and clutch as an assembly, first drain cooling system and if engine is to be disassembled, drain oil pan. Remove the hood and grille, then disconnect radiator hoses, head light wires and the radiator upper support bracket. Unbolt and remove the radiator and shell assembly from bolster. Disconnect the radius rod pivot bracket from clutch housing and both drag links from the steering (knuckle) arms. Support tractor under clutch housing, unbolt the bolster support brackets from engine and roll bolster, wheels and axle assembly forward and away from tractor.

Disconnect cable and remove starting motor. Disconnect the heat indicator sending unit, hydraulic lines, fuel lines, oil pressure gage line, wiring harness and controls from engine and engine accessories. On Diesel models, remove the final fuel filter and air cleaner hose.

Swing engine in a hoist, unbolt engine from clutch housing and move engine forward and away from tractor. Note: The two long bolts retaining top of clutch housing to engine are unscrewed gradually as engine is moved forward. This procedure eliminates the need of removing the steering gear unit and usually saves considerable time.

Series 350D

30A. To remove the engine and clutch as an assembly, first remove hood; then on models with manual steering, disconnect the steering shaft universal joint and remove the U-joint Woodruff key. On models with power steering, loosen the steering shaft coupling which is located in front of the power unit. Disconnect the radiator upper support bracket, drain cooling system and disconnect the radiator hoses. If engine is to be disassembled, drain oil pan. Remove the dust shield from under front end of frame

rails and on models with an adjustable type front axle, disconnect the radius rod at its rear ball joint. Support tractor under clutch housing, unbolt the upper bolster from the frame rails and move the bolster, radiator and front wheels assembly away from tractor.

Remove air cleaner and hose. Disconnect the heat indicator sending unit, fuel lines, oil pressure gage line, wiring harness and controls from engine and engine accessories. Remove the final fuel filter and disconnect the hydraulic oil lines on right side of tractor.

Attach hoist to engine in a suitable manner and remove the frame side rails. Unbolt engine from clutch housing and move engine and clutch assembly forward and away from tractor.

CYLINDER HEAD

Series 300-350-400-450

31. To remove the cylinder head, first drain cooling system and remove hood. On models so equipped, remove the radiator upper support rod.

On models with manual steering, remove the grille and the steering worm shaft.

On models with power steering, remove the grille, loosen the steering shaft coupling located in front of the steering power unit and slide the coupling forward on the worm shaft. Remove the steering worm shaft front bearing retainer and turn the worm shaft forward and out of the steering gear housing. Remove the air cleaner and remove the two strap iron braces retaining the power unit to the fuel tank support. Loosen the clamp bolt in the steering shaft U-joint, slide the power unit forward and remove the U-joint Woodruff key. Then, slide the unit forward and, without disconnecting the oil lines, lay the unit rearward and out of way.

On all models, remove the heat indicator sending unit, generator and bracket and the water outlet casting. Disconnect the fuel lines and carburetor controls. Remove the valve cover, rocker arms assembly and push rods. Remove the cylinder head retaining stud nuts and lift cylinder head from tractor.

When installing the cylinder head, tighten the stud nuts evenly and to a torque of 70 ft.-lbs. for the 300 & 350 series; 110 ft.-lbs. for the 400 & 450 series.

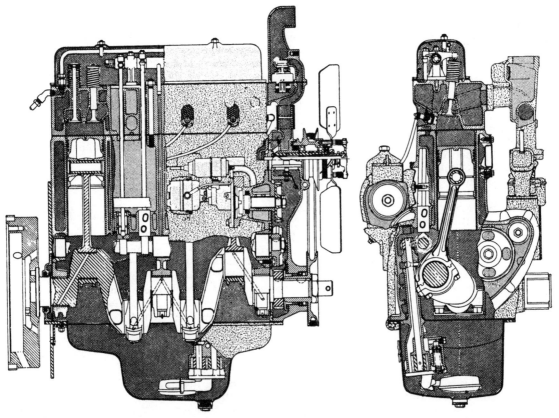

Fig. IH615—Side and end sectional views of the 264 cubic inch non-Diesel engine. Other non-Diesel engines are similarly constructed. The unit shown is equipped with battery ignition; magneto ignition, however, is optionally available on some models.

Series 300U-350U-W400-W450

31A. To remove the cylinder head, first drain cooling system, remove the hood and on series W400 & W450, remove the radiator brace rod. On all models, remove heat indicator sending unit, generator and bracket and water outlet casting. Disconnect fuel lines and carburetor controls.

Remove valve cover, rocker arms assembly and push rods. Remove cylinder head retaining stud nuts and lift cylinder head from tractor.

When installing cylinder head, tighten the stud nuts evenly and to a torque of 70 ft.-lbs. for the 300U & 350U series; 110 ft.-lbs. for the W400 & W450 series.

Series 400D-450D

32. To remove the cylinder head, first drain cooling system and remove the hood and radiator upper support rod.

On models with manual steering, remove the hood and the steering worm shaft.

On models with power steering, remove the grille, loosen the steering shaft coupling located in front of the steering power unit and slide the coupling forward on the worm shaft.

Remove the steering worm shaft front bearing retainer and turn the worm shaft forward and out of the steering gear housing. Remove the air cleaner and remove the two strap iron braces retaining the power unit to the fuel tank support. Loosen the clamp bolt in the steering shaft U-joint, slide the power unit forward and remove the U-joint Woodruff key. Then, slide the unit forward and, without disconnecting the oil lines, lay the unit rearward and out of way.

On all models, remove the heat indicator sending unit, water outlet casting, injection pump to nozzle lines and carburetor air cleaner pipe. Disconnect governor arm and starting linkage and remove the center steering shaft guide from bracket. Disconnect wiring harness and remove the carburetor fuel lines and controls. Remove valve cover, rocker arms assembly and push rods. Remove the cylinder head retaining stud nuts and using a hoist, lift cylinder head from tractor.

Before installing the cylinder head, examine both expansion plugs which are located in bottom side of head. Always renew a questionable plug. When installing the head, tighten the stud nuts evenly and to a torque of 110-135 ft.-lbs.

Series W400D-W450D

32A. To remove the cylinder head, first drain cooling system and remove hood and radiator brace rod. Remove the heat indicator sending unit, water outlet casting, injection pump-to-nozzle lines and carburetor air cleaner pipe. Disconnect governor arm and starting linkage. Disconnect wiring harness and remove fuel lines and controls from carburetor. Remove valve cover, rocker arms assembly and push rods. Remove the cylinder head retaining stud nuts and using a hoist, lift cylinder head from tractor.

Before installing the cylinder head, examine both expansion plugs which are located in bottom side of head. Always renew a questionable plug. When installing head, tighten the stud nuts evenly and to a torque of 110-135 ft.-lbs.

Series 350D

32B. To remove the cylinder head, first drain cooling system, remove hood and disconnect the radiator upper support bracket.

On models with manual steering, remove the grille and the steering worm shaft.

On models with power steering, remove the grille, loosen the steering

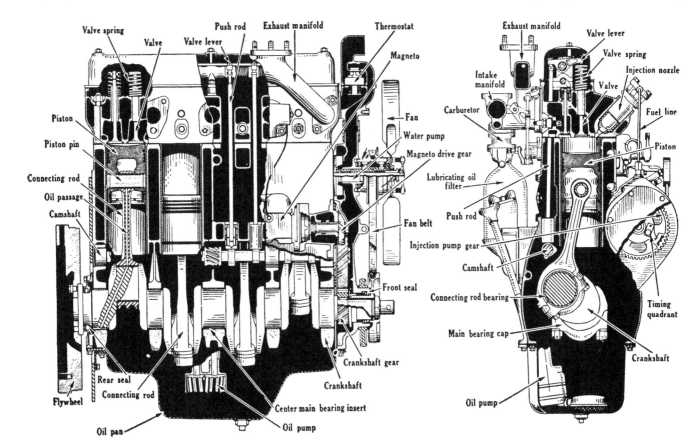

Fig. IH616—Side and end sectional views of the 264 cubic inch engine used in series 400D and W400D. The 281 cubic inch engine used in series 450D and W450D is similar. The Diesel injection pumps and nozzles are manufactured by International Harvester.

shaft coupling located in front of the steering power unit and slide the coupling forward on the worm shaft. Remove the steering worm shaft front bearing retainer and turn the worm shaft forward and out of the steering gear housing. Remove the air cleaner and remove the strap iron brace retaining the power unit to the intake manifold. Loosen the clamp bolt in the steering shaft U-joint and unbolt steering shaft center bearing from fuel tank support. Then, slide the unit forward and, without disconnecting the oil lines, lay the unit rearward and out of way.

On all models, remove the air cleaner and hose, heat indicator sending unit, water outlet casting and disconnect the interferring controls, wiring harness and hydraulic lines.

Remove the high pressure lines connecting the injection pump to the injection nozzles and disconnect the leak-off line. Immediately cover the connections with composition caps or tape to prevent the entrance of dirt. Remove the valve cover, rocker arms assembly and push rods. Remove the inlet and exhaust manifold units.

Remove the cylinder head retainer stud nuts and lift the cylinder head from tractor.

When installing the cylinder head, tighten the stud nuts evenly and to a torque of 35-40 ft.-lbs. for 3/8-inch studs, 100-110 ft.-lbs. for 1/2-inch studs. Manifold bolts should be tightened to a torque of 25-30 ft.-lbs.

Series 350DU

32C. To remove the cylinder head, first drain cooling system, remove hood and disconnect the radiator upper support bracket. Remove the heat

indicator sending unit, water outlet casting and disconnect the interferring controls, wiring harness and hydraulic lines.

Remove the high pressure lines connecting the injection pump to the injection nozzles and disconnect the leak-off line. Immediately cover the connections with composition caps or tape to prevent the entrance of dirt. Remove the valve cover, rocker arms assembly and push rods. Remove the inlet and exhaust manifold units.

Remove the cylinder head retaining stud nuts and lift the cylinder head from tractor.

When installing the cylinder head, tighten the stud nuts evenly and to a torque of 35-40 ft.-lbs. for 3/8-inch studs, 100-110 ft.-lbs. for 1/2-inch studs. Manifold bolts should be tightened to a torque of 25-30 ft.-lbs.

VALVES AND SEATS
Series 300-300U-350-350U

33. Intake and exhaust valves are not interchangeable. Intake valves seat directly in cylinder head; whereas, the cylinder head is fitted with renewable seat inserts for the exhaust valves. Valves have a seat angle of 45 degrees and a seat width of $\frac{1}{16}$-inch. Seats can be narrowed, using 15 and 70 degree stones. Valves used in kerosene and distillate models are equipped with stem retainers to prevent valve from dropping into combustion chamber. Valves have a stem diameter of 0.3405-0.3415. Tappet gap should be set Hot to 0.017 for both intake and exhaust.

Series 400-W400-450-W450

34. Intake and exhaust valves are not interchangeable. Intake valves seat directly in cylinder head; whereas, the cylinder head is fitted with renewable seat inserts for the exhaust valves. Valves have a seat angle of 45 degrees and it is recommended that the face angle be finished to 44-44¼ degrees.

Intake and exhaust valve seat width is 5/64-inch. Seats can be narrowed, using 15 and 70 degree stones. Valves are equipped with safety stem retainers to prevent valve from dropping into combustion chamber. Valves have a stem diameter of 0.371-0.372. Tappet gap should be set to 0.017 Hot.

Series 400D-W400D-450D-W450D

35. RUNNING VALVES. Intake and exhaust valves are not interchangeable, and seat directly in cylinder head with a seat angle of 45 degrees and a seat width of $\frac{3}{32}$-inch. Seats can be narrowed, using 15 and 70 degree stones. Valve stem diameter is 0.371-0.372. Tappet gap should be set to 0.017 Hot.

36. STARTING VALVES. Starting valves seat directly in cylinder head with a seat angle of 45 degrees and a seat width 3/64-inch. Valves have a stem diameter of 0.309.

Before installing the valve cover, refer to Fig. IH617 and check the following: Place the starting controls in the Diesel running position and check to make certain that there is some clearance between each of the starting valve covers and the valve actuating arms on the starting valve shaft. Switch the controls to the gasoline starting position; at which time, the starting valve shaft arms have opened the starting valves. Then, using a screw driver as shown in Fig. IH617, push down on the starting valve covers to make certain that each cover has an additional downward travel of at least 1/64-inch.

If the preceeding checks are not as specified, the starting linkage should be adjusted as outlined in paragraph 111.

Series 350D-350DU

36A. Intake and exhaust valves are not interchangeable. Intake valves seat directly in cylinder head; whereas, the cylinder head is fitted with renewable seat inserts for the exhaust valves. Valves have a seat angle of 45 degrees and a seat width of $\frac{1}{16}-\frac{3}{32}$-inch. Seats can be narrowed, using 15 and 70 degree stones. Valves have a stem diameter of 0.3406-0.3414 for the intake, 0.3382-0.3390 for the exhaust. Tappet gap should be set Hot to 0.014.

36B. Valve seat inserts can be removed with a suitable puller. New inserts are available in 0.010 oversize. Machine the inside diameter of the counterbore 0.003-0.005 smaller than

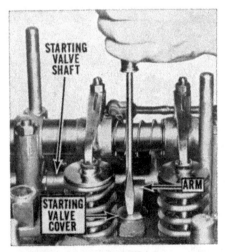

Fig. IH617 — Checking the starting valve cover travel on series 400, W400, 450 & W450 Diesel engines. With the controls in the starting position, the covers should have an additional downward travel of 1/64-inch when manually pushed down as shown.

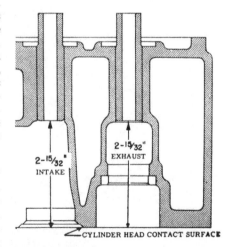

Fig. IH617A—Sectional view of cylinder head used on series 350D and 350DU, showing the valve guide installation dimensions. Intake and exhaust valve guides are interchangeable.

the outside diameter of the new insert. This will provide approximately 0.004 interference fit. Chill new insert in dry ice to facilitate installation. Make certain that insert is firmly seated in counterbore, then roll or peen in place.

VALVE GUIDES AND SPRINGS
Series 350D-350DU

36C. Intake and exhaust valve guides are interchangeable and can be pressed from cylinder head if renewal is required. Press new guides into cylinder head from combustion chamber side until the distance from port end of guide to gasket surface of head is $2\frac{1}{32}$ inches as shown in Fig. IH617A. New guides are pre-sized and if carefully installed, will require no reaming.

Valve stem to guide clearance is 0.0008-0.0026 for the intake, 0.0032-0.0050 for the exhaust.

36D. Intake and exhaust valve springs are interchangeable. Springs should have a free length of 2⅜ inches and should test 115-123 pounds when compressed to a height of 1.521 inches. Renew any spring which is rusted, discolored or does not meet the pressure test specifications. Springs should be installed with closely wound coils next to cylinder head.

Series 300-300U-350-350U-400-400D-W400-W400D-450-450D-W450-W450D

37. Intake and exhaust valve guides are interchangeable in any one model. On Diesel models, however, the starting valve guides are not interchangeable with the running valve guides. Before removing worn guides from cylinder head, measure the distance from top of guide to valve spring seating surface on top of head and install new guides to the same dimension.

Valve guides are pre-sized, and if carefully installed, will require no

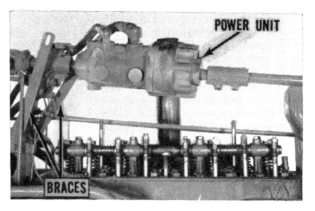

Fig. IH 617C—On series 300, 350, 400 and 450, with power steering, the power steering unit must be turned upward as shown before the rocker arm cover can be removed.

final sizing. Recommended valve stem to guide clearance is 0.002-0.004.

38. Intake and exhaust valve springs are interchangeable in any one model. On Diesel models, however, the starting valve springs are not interchangeable with running valve springs.

Valve springs with damper coils (closely wound coils at one end) should be installed with closed coils next to the cylinder head. Renew any spring which is rusted, discolored or does not meet the pressure test specifications which follow:

Valve Spring Free Length, (Inches):
Series 300-300U-350-350U 2 9/32

Series 400-W400-450-W450 ...2 37/64

Series 400D-W400D-450D-W450D (Running Springs)..2 11/32

Series 400D-W400D-450D-W450D (Starting Springs)..1 31/32

Test load (Lbs.) @ Length, (Inches):
Series 300-300U-350-350U ....81-89 @ 1 25/64

Series 400-W400-450-W450 ....106 @ 1 35/64

Series 400D-W400D-450D-W450D (Running Springs) ...147 @ 1 9/16

Series 400D-W400D-450D-W450D (Starting Springs) .....24 @ 1 5/32

VALVE TAPPETS (CAM FOLLOWERS)
Series 300-300U-350-350U-400-400D-W400-W400D-450-450D-W450-W450D

39. Tappets are of the ported barrel type and ride directly in bores cut in the crankcase (cylinder block). Clearance of tappets in crankcase bores should not exceed the I&T suggested limit of 0.005. Oversize tappets are not available. Tappets are removed from the side of crankcase after removing valve levers (rocker arms) assembly, push rods, side cover plate (or plates) and tappet stop.

Series 350D-350DU

39A. The mushroom type tappets (cam followers) ride directly in the cylinder block bores and are available in standard size only. Clearance of tappets in crankcase bores should not exceed the I&T suggested limit of 0.005. To remove the tappets, it is first necessary to remove the camshaft as outlined in paragraph 52A.

VALVE TAPPET LEVERS (ROCKER ARMS)
Series 350D-350DU

39B. The procedure for removing the rocker arms and shaft assembly on series 350DU and series 350D with manual steering is conventional. On series 350D with power steering, unbolt the bracket retaining the steering motor to the intake manifold and turn the steering motor enough to permit removal of the valve cover.

Tappet lever shaft and/or rocker arms should be renewed if clearance between the shaft and rocker arms exceeds 0.005.

Series 300-300U-350-350U-400-400D-W400-W400D-450-450D-W450-W450D

40. REMOVE AND REINSTALL. The procedure for removing the rocker arms assembly on series 300U, 350U, W400, W400D, W450 and W450D is

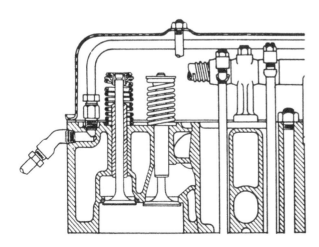

Fig. IH617B — Sectional view of gasoline engine cylinder head showing installation details of valves, guides and springs. A sectional view of the valve rotator is shown in Fig. IH618.

conventional and readily evident after an examination of the unit.

On series 300, 350, 400 & 450 with manual steering, the procedure for removing the rocker arms assembly is conventional, but on models with power steering, proceed as follows: Remove the hood sections and unbolt the two strap iron braces from the power steering unit. Turn the unit until the hose connections are up; then remove the rocker arm cover as shown in Fig. IH617C. The remaining procedure is conventional.

On series 400D & 450D with manual steering, the procedure for removing the rocker arms assembly is conventional, but on models with power steering, it is first necessary to remove the power steering unit as outlined in paragraph 23 before the rocker arm cover and arms can be removed.

40A. **OVERHAUL.** On above non-Diesel models, the valve levers and hollow shaft assembly is pressure lubricated from the center camshaft bearing via an oil sleeve which is mounted on the valve lever shaft. When installing the valve levers and shaft assembly, the tube on the oiler sleeve must enter the reamed hole in the cylinder head. The valve lever shaft diameter is 0.748-0.749.

41. On Diesel models, the valve levers and hollow shaft assembly is pressure lubricated from the camshaft front bearing via an oil passage in the cylinder head. The valve lever shaft diameter is 0.872-0.873.

42. Some models are equipped with forged type valve levers which are equipped with renewable type bushings. These busings require final sizing after installation to provide a clearance of 0.002-0.004 for the valve lever shaft. When installing the bushings, make certain that the oil hole in bushing is in register with oil spurt hole in the valve lever. On other models, the valve levers are of the welded construction with a non-serv-

iceable bushing. The welded type lever which is interchangeable with the forged type on non-Diesels, should be renewed whenever the lever-to-shaft clearance exceeds the I&T suggested limit of 0.008.

VALVE ROTATORS
All Models So Equipped

43. Positive type valve rotator (Rotocaps) are installed on some models.

Normal servicing of the valve rotators consists of renewing the units. It is important, however, to observe the valve action after engine is started. The valve rotator action can be considered satisfactory if the valve rotates a slight amount each time the valve opens. A cut-away view of a typical "Rotocap"installation is shown in Fig. IH618. Valve spring test specifications are listed in paragraph 38.

VALVE TIMING
All Models

44. Valves are properly timed when the punch marked timing gear teeth are meshed as outlined in paragraphs 49, 50 and 50A.

TIMING GEAR COVER
Series 300-350-400-400D-450-450D

45. To remove the timing gear cover, first remove hood and on models with manual steering, disconnect the steering shaft universal joint and remove the U-joint Woodruff key. On models with power steering, loosen the steering shaft coupling which is located in front of the power unit. Disconnect the radiator upper support bracket or rod, drain cooling system and disconnect the radiator hoses. Disconnect the shutter control rod. Remove the dust shield from under

front end of frame rails and on models with an adjustable type front axle, disconnect the radius rod at its rear ball joint. Support tractor under clutch housing, unbolt the upper bolster from the frame rails and move the bolster, radiator and front wheels assembly away from tractor.

Remove the fan blades and crankshaft pulley. On Diesel models, remove the generator, place floor jack under engine and remove the engine front supporting cross member.

On all models, unbolt and remove cover from engine. The lip type crankshaft front oil seal can be renewed at this time.

When reinstalling the cover, leave the retaining cap screws loose until after the crankshaft pulley is installed and the crankshaft turned several revolutions. This procedure will facilitate centering the front oil seal with respect to the pulley hub.

Series W400-W400D-W450-W450D

46. To remove the timing gear cover, first remove the hood and the radiator upper support rod. Disconnect the shutter control rod, drain cooling system and disconnect the radiator hoses. Disconnect the steering drag link from steering knuckle arm and the radius rod at its rear ball joint. Disconnect the head light wires. Support tractor under clutch housing, unbolt the upper bolster from the frame rails and move the bolster, radiator and front wheels assembly away from tractor.

Remove the fan blades and crankshaft pulley. On Diesel models, remove the generator, place floor jack under engine and remove the engine front supporting cross member.

On all models, unbolt and remove cover from engine. The lip type crankshaft front oil seal can be renewed at this time.

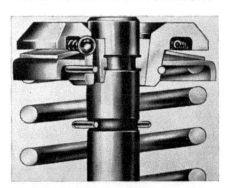

Fig. IH618—Sectional view of a typical installation of a valve rotator ("Rotocap") on models so equipped.

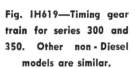

Fig. IH619—Timing gear train for series 300 and 350. Other non-Diesel models are similar.

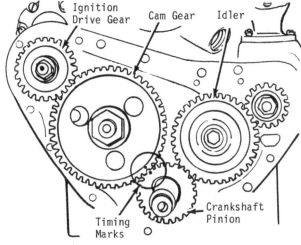

When reinstalling the cover, leave the retaining cap screws loose until after the crankshaft pulley is installed and the crankshaft turned several revolutions. This procedure will facilitate centering the front oil seal with respect to the pulley hub.

Series 300U-350U-350DU

47. To remove the timing gear cover, first drain cooling system and remove hood and grille. Disconnect the radiator hoses, head light wires and the radiator upper support bracket. Unbolt and remove the radiator and shell assembly from bolster. Disconnect the radius rod pivot bracket from clutch housing and both drag links from the steering knuckle arms. Support tractor under clutch housing, unbolt the bolster support brackets from engine and roll the bolster, wheels and axle assembly forward and away from tractor.

Remove the fan belt and crankshaft pulley. Remove the front two oil pan retaining cap screws and loosen the others. On Diesel models, remove the hydraulic pump and adaptor assembly. On all models unbolt and remove timing gear cover from engine, using care not to damage the front of the oil pan gasket which may adhere to the timing gear cover. The lip type crankshaft front oil seal can be renewed at this time.

Series 350D

47A. To remove the timing gear cover, first remove hood and grille and on models with manual steering, disconnect the steering shaft universal joint and remove the U-joint Woodruff key. On models with power steer-ing, loosen the steering shaft coupling which is located in front of the power unit. On all models remove worm-shaft. Disconnect the radiator upper support bracket, drain cooling system and disconnect the radiator hoses. Remove the dust shield from under front end of frame rails and remove radiator. Support tractor under clutch housing, and remove the left frame rail.

Remove the fan blades, belt and crankshaft pulley. Remove the front two oil pan retaining cap screws and loosen the others. Remove the hydraulic pump and adapter assembly. Unbolt the timing gear cover from crankcase, carefully separate the oil pan gasket from the timing gear cover and remove cover from engine as shown in Fig. IH619B. The crankshaft front oil seal can be renewed at this time.

Install the cover by reversing the removal procedure and use sealer on all gasket surfaces.

TIMING GEARS
Series 300-300U-350-350U-400-400D-W400-W400D-450-450D-W450-W450D

48. The camshaft gear, crankshaft gear and/or idler gear can be renewed after removing the crankcase front cover as outlined in paragraph 45, 46 or 47. The idler gear which on non-Diesels contains a bushing, rotates on a shaft which is bolted to front of crankcase. Clearance of the bushing on the shaft should be 0.001-0.0025. The idler gear on Diesel models is fitted with an anti-friction bearing.

49. When installing the timing gears on non-Diesels, mesh single punch marked tooth on crankshaft gear with the single punch marked tooth space on camshaft gear; then mesh double punch marked tooth space on camshaft gear with the double punch marked tooth on the ignition unit drive gear. The governor gear and idler gear need not be timed to the engine.

50. When installing the timing gears on Diesels, mesh single punch mark on crankshaft gear with single punch mark on idler gear; mesh triple punch mark on idler gear with triple punch mark on the injection pump drive gear; mesh double punch mark on idler gear with double punch mark on camshaft gear; and mesh single punch mark on camshaft gear with single punch mark on magneto or distributor drive gear.

Series 350D-350DU

50A. The camshaft gear and/or crankshaft gear can be removed by using suitable pullers after first removing the timing gear cover as outlined in paragraph 47 or 47A.

When installing the timing gears, mesh the single punch marked tooth on crankshaft gear with the double punch marked tooth space on camshaft gear. Tighten the camshaft gear retaining nut to a torque of 115-125 ft.-lbs.

CAMSHAFT
Series 300-300U-350-350U-400-400D-W400-W400D-450-450D-W450-W450D

51. To remove the camshaft, first remove the crankcase front cover as outlined in paragraph 45, 46 or 47. Remove oil pan, oil pump and valve levers (rocker arms) assembly. Refer to paragraph 40. Remove tappet cover, tappet stop and lift tappets from their bores in crankcase. Pull camshaft

Fig. IH619A—Timing gear train for series 350D. Notice that the punch marked tooth on crankshaft gear is meshed with the punch marked tooth space on camshaft gear.

Fig. IH619B—Timing gear cover can be removed from series 350D engines after radiator and left side rail are removed as shown. Some mechanics prefer to remove the complete bolster and front end unit.

gear, remove the camshaft thrust plate retaining cap screws and withdraw camshaft from front of engine.

52. Normal camshaft end play of 0.003-0.011 for non-Diesels and 0.005-0.013 for Diesels is controlled by a thrust plate which is located between the camshaft gear and the crankcase. Excessive camshaft end play is corrected by renewal of the thrust plate. The camshaft is carried in steel-backed, babbitt-lined bushings. Renewal of the bushings requires removal of engine from tractor. If bushings are pressed or driven into position with a piloted arbor which is a close fit in the bushings, they will not require reaming (check bushings for high spots after installation by installing camshaft). It may be necessary to scrape the bearings if bore is irregular. When installing bushings, the forward end of each bushing should be flush with front face of its bore and oil holes must be lined up with oil passages.

Check camshaft against the values listed below:

No. 1 (front) journal diameter:
  Series 300-300U-
    350-350U 1.9305-1.9315
  Series 400-W400-
    450-W450 . . . . .2.243 -2.244
  Series 400D-W400D-
    450D-W450D 2.4305-2.4315

No. 2 journal diameter:
  Series 300-300U-
    350-350U 1.8055-1.8065
  Series 400-W400-
    450-W450 . . . . .2.118 -2.119
  Series 400D-W400D-
    450D-W450D 2.3055-2.3065

No. 3 journal diameter:
  Series 300-300U-
    350-350U . . . . .1.368 -1.369
  Series 400-W400-
    450-W450 . . . . .1.868 -1.869
  Series 400D-W400D-
    450D-W450D 2.1805-2.1815

No. 4 journal diameter:
  Series 400D-W400D-
    450D-W450D . . . . . . .1.868 -1.869
Journal running
  clearance 0.0015-0.0035
Camshaft end play
  Non-diesels . . . . .0.003 -0.011
Camshaft end play
  diesels . . . . .0.005 -0.013

Oil leakage around rear of the camshaft is prevented by an expansion plug. Renewal of this plug is accomplished by first removing flywheel.

Series 350D-350DU

52A. To remove the camshaft, first remove the crankcase front cover as outlined in paragraph 47 or 47A. Remove the oil pan, injection pump, injection pump drive gear, rocker arms assembly and push rods. Using a suitable puller, remove the camshaft timing gear. Remove the camshaft thrust plate retaining cap screws, push the tappets up and withdraw camshaft from crankcase. Check the camshaft bearing journals against the following values:

No. 1 (front) . . . . .1.808 -1.809
No. 2 . . . . .1.7455-1.7465
No. 3 . . . . .1.683 -1.684

Recommended camshaft end play of 0.005-0.009 is controlled by the thickness of the thrust plate which is located between the camshaft gear and crankcase. Excessive camshaft end play is corrected by renewal of the thrust plate.

The camshaft is carried in three steel-backed, babbitt-lined bushings. Renewal of the bushings requires removal of engine from tractor. If bushings are pressed or driven into position with a piloted arbor which is a close fit in the bushings, they will not require reaming. Recommended clearance of camshaft in the installed bushings is 0.003-0.006. When installing the bushings, make certain that the bushing oil holes register with the oil feed holes in the crankcase.

Oil leakage around rear of the camshaft is prevented by an expansion plug. Renewal of this plug requires R&R of flywheel.

ROD AND PISTON UNITS
All Models

53. Connecting rod and piston units are removed from above after removing the cylinder head and oil pan. Cylinder numbers are stamped on connecting rod and cap. When installing the connecting rod and piston units, make certain that numbers on rods and caps are in register and face toward camshaft side of engine. Tighten the connecting rod bolts to the torque values which follow:

Series 300-300U-
  350-350U . . . . 40 ft.-lbs.
Series 350D-350DU . . . .70-75 ft.-lbs.
Series 400-W400-
  450-W450 . . . . .55-60 ft.-lbs.
Series 400D-W400D-
  450D-W450D . . . . .115 ft.-lbs.

PISTONS, SLEEVES AND RINGS
Series 300-300U-350-350U-400-W400-450-W450

54. Pistons are available for standard compression ratio engines, for special compression ratio engines and engines for operation at 5000 and 8000 foot altitudes. Pistons are available in matched units with the sleeves. The matched units are available individually or in sets of four.

Note: Lightweight pistons, connecting rods, piston pins, piston pin bushings and retainer rings are regular equipment on the gasoline burning standard altitude series 300, 300U, 400 and W400, as follows — F300 after 25079, F300HC after 25091, F300L after 25091, I300 after 23300, F400 after 31238, F400L after 31258, I400 after 2483, F400 HC after 31259 and I400 L after 2484. Lightweight pistons must be used with lightweight connecting rods as the piston pins used with lightweight pistons are smaller.

EDITOR'S NOTE: At the time of this writing, lightweight piston and rod assemblies were available for standard altitude engines only. When servicing high altitude engines, the local IH parts department should be contacted regarding the availability of light weight units for service installation.

Recommended piston skirt clearance is as follows:
Series 300-300U
  (heavyweight) . . . . .0.003 -0.004
Series 300-300U
  (lightweight) . . . . . .0.002 -0.003
Series 350-350U . . . . . . .0.002 -0.003
Series 400-W400
  (heavyweight) 0.0029-0.0039
Series 400-W400
  (lightweight) 0.0025-0.0035
Series 450-W450 . . . . . .0.0025-0.0035

55. The pistons and sleeves should be renewed if wear exceeds the values which follow:
Sleeve out-of-round
  Series 300-300U-
    350-350U 0.0035
  Series 400-W400-
    450-W450 . . . . .0.004
Sleeve taper
  Series 300-300U-
    350-350U 0.011
  Series 400-W400-
    450-W450 . . . . .0.012

56. To renew the dry type sleeves after pistons are out, remove them from above using IH puller SE 1213 or equivalent. Thoroughly clean crankcase bores and the counterbore at top and clean any paint or grease from the sleeves. Coat the outside of sleeves with light engine oil and press or drive sleeves into bores of crankcase. Tops of sleeves should be flush with

top of crankcase but may project above it a maximum of 0.006. Top of sleeve should NOT be below top surface of crankcase. Cylinder sleeves do not require honing or boring after installation but should be checked for possible distortion or localized high spots.

57. Check the piston rings against the values which follow:

End gap
  Series 300-300U-
    350-350U ....0.010 -0.018
  Series 400-W400-
    450-W450 ....0.013 -0.023
Top compression ring
  side clearance 0.0035-0.005
Other compression ring
  side clearance ....0.002 -0.0035
Oil ring side clearance...0.0015-0.003

Rings stamped "Top", should be installed with word top facing toward top of piston. Rings having a groove in the outer face should be installed with groove down.

Series 400D-W400D-450D-W450D

58. Pistons and/or sleeves are available separately or in matched sets. Recommended piston skirt clearance is 0.0046-0.0054. Install pistons with word "Front" toward front of engine.

59. The dry type cylinder sleeves should be renewed if the out-of-round exceeds 0.0037 or if the taper exceeds 0.0155.

Sleeve stand-out above cylinder block is 0.039-0.047. The procedure for renewing the cylinder sleeves is outlined in paragraph 56.

60. Check the piston rings against the values which follow:

End gap 0.012-0.028
Top comp. ring side clearance...0.005
Other ring side clearance......0.003

Rings stamped "Top", should be installed with word top facing toward top of engine.

Series 350D-350DU

60A. Each aluminum alloy piston is fitted with three 1/8-inch wide compression rings and one 1/4-inch wide oil control ring. Piston ring end gap should be 0.010-0.018. Side clearance in piston grooves should be 0.003-0.0045 for the compression rings, 0.0015-0.0035 for the oil control ring. Pistons should have a skirt clearance of 0.0045-0.0055 in the dry type cylinder sleeves. Re-ring sets should correct excessive oil consumption providing the sleeve taper does not exceed 0.008. If, however, the sleeve taper exceeds 0.008, matched sets of pistons and sleeves should be installed. Pis-

tons and sleeves are available as matched units and are stamped with size classification marks such as A, B, C, etc. When reassembling, be sure to install the proper piston in the proper sleeve.

Sleeves can be removed, using a suitable puller. When reassembling, coat outside of sleeves with light oil, start the sleeves in their respective bores with the connecting rod reliefs 90 degrees to crankshaft and press the sleeves into crankcase. If undue pressure is required to install sleeves, remove them and check for foreign material.

Tapered piston rings must be installed with word "TOP" toward top of engine.

PISTON PINS
Series 350D-350DU

61. The full floating type piston pins are available in standard size only. Piston pin diameter is 1.1091-1.1093. Piston pin should have a light push fit in piston when piston is heated to 160 degrees F. Pins should have a clearance of 0.0002-0.0006 in connecting rod bushing.

Series 300-300U-350-350U-400-400D-W400-W400D-450-450D-W450-W450D

61A. The full floating type piston pins are available in standard and 0.005 oversize. Check piston pin against the values which follow:

Piston pin diameter:
  Series 300-300U
    (heavyweight) 1.1089-1.1092
  Series 300-300U
    (lightweight) 0.8748-0.8749
  Series 350-350U 0.8748-0.8749
  Series 400-400D-W400-W400D
    (heavyweight) 1.3125-1.3127
  Series 400-W400
    (lightweight) 0.9998-0.9999
  Series 450-W450 ......0.9998-0.9999
  Series 450D-W450D ...1.3125-1.3127
Piston pin clearance
  in piston 0.0000-0.0003
Piston pin clearance in rod bushing:
  Non-diesels 0.0002-0.0005
  Series 400D-W400D ...0.003 -0.005
  Series 450D-W450D ...0.0004-0.0009

CONNECTING RODS AND BEARINGS
Series 350D-350DU

62. Connecting rod bearings are of the slip-in, precision type, renewable from below after removing the oil

pan and rod bearing caps. Bearing inserts are available in standard size only. Check the crankshaft and rod bearing inserts against the values which follow:

  Crankpin diameter ...2.249 -2.250
  Bearing running
    clearance 0.0006-0.0031
  Rod side play........ 0.006
  Rod bolt torque......70-75 ft.-lbs.

Series 300-300U-350-350U-400-400D-W400-W400D-450-450D-W450-W450D

62A. Connecting rod bearings are of the slip-in, precision type, renewable from below after removing oil pan and rod bearing caps. When installing new bearing shells, make certain that the rod and bearing cap numbers are in register and face toward camshaft side of engine. Bearing inserts are available in standard size as well as undersizes of 0.002, 0.003, 0.030 and 0.032 for non-diesels, 0.002, 0.030 and 0.032 for diesels. Check the crankshaft and bearing inserts against the values which follow:

Crankpin diameter:
  Series 300-300U-
    350-350U 2.2975-2.2985
  Series 400-W400-
    450-W450 ....2.5475-2.5485
  Series 400D-W400D-
    450D-W450D 3.2475-3.2485
Running clearance:
  Non-diesels 0.0011-0.0037
  Diesels ....0.002 -0.003
Side play:
  Non-diesels ....0.005 -0.012
  Diesels ....0.003 -0.010
Rod bolt torque: (ft. lbs.)
  Series 300-300U-350-350U....... 40
  Series 400-W400-450-W450....55-60
  Series 400D-W400D-
    450D-W450D 115

CRANKSHAFT AND MAIN BEARINGS
Series 350D-350DU

63. The crankshaft is supported in three slip-in, precision type main bearings renewable from below after removing the oil pan, oil pump and main bearing caps. Main bearings are available in standard size only.

Normal crankshaft end play of 0.004-0.008 is controlled by the flanged center main bearing. Excessive end play is corrected by renewing the center main bearing shells. The International Harvester Co. specifies that crankshaft should not be ground to undersize. If bearing journals are out of round more than 0.0015, the crankshaft should be renewed. To remove the crankshaft, remove engine, clutch, flywheel, rear oil seal retainer, timing gear cover, oil pump, rod and main bearing caps.

Check the crankshaft and main bearings against the values which follow:

Crankpin
   diameter...Refer to paragraph 62.
Main journal diameter.2.374 -2.375
Main bearing running
   clearance 0.0009-0.0036
Crankshaft end play...0.004 -0.008
Main bearing
   bolt torque ....100-110 ft.-lbs.

Refer to paragraph 66 when installing the oil pump.

Series 300-300U-350-350U-400-400D-W400-W400D-450-450D-W450-W450D

63A. The crankshaft is supported in three main bearings on non-Diesel engines, five main bearings on diesels. End thrust is taken on the center main bearing. Main bearings are of the shimless, non-adjustable, slip-in precision type, renewable from below after removing the oil pan and main bearing caps. Removal of the rear main bearing cap on all models, requires removal of the crankshaft rear oil seal lower retainer plate. Renewal of crankshaft requires R&R of engine. Check crankshaft and main bearings against the values which follow:

Crankpin
   diameter ....Refer to paragraph 62.
Main journal diameter:
   Series 300-300U-
      350-350U 2.5575-2.5585

Series 400-W400-
   450-W450 ....2.8075-2.8085
Series 400D-W400D-
   450D-W450D 3.7475-3.7485
Crankshaft end-play ....0.004 -0.008
Main bearing running clearance:
   Series 300-300U-
      350-350U 0.0011-0.0037
   Series 400-W400-
      450-W450 ....0.002 -0.003
   Series 400D-W400D-
      450D-W450D 0.0018-0.0048
Main bearing bolt torque, (ft. lbs.):
   Series 300-300U-
      350-350U 75
   Series 400-W400-
      450-W450 100-105
   Series 400D-W400D-
      450D-W450D (⅝-studs) ..150-175
   Series 400D-W400D-
      450D-W450D (¾-studs) ..250-275
Main bearings are available in standard size as well as undersizes of 0.002, 0.003, 0.030 and 0.032 for non-diesels, 0.002, 0.030 and 0.032 for diesels.

CRANKSHAFT REAR OIL SEAL
All Models

64. **RENEW.** Procedure for renewal of the crankshaft rear oil seal is evident after removing the flywheel as outlined in paragraph 65.

FLYWHEEL
All Models

65. The flywheel can be removed after splitting engine from clutch housing and removing the clutch.

To install the flywheel ring gear, heat same to approximately 500 deg. F. and install gear on flywheel so that beveled end of the ring gear teeth will face beveled end of teeth on starting motor pinion.

OIL PUMP
Series 350D-350DU

66. The gear type oil pump is located internally, under the crankshaft front main bearing cap. The unit is driven from the crankshaft timing gear. Pump removal procedure is evident after removing the oil pan.

Refer to Fig. IH619C, disassemble the pump and renew any parts which are worn, scored or damaged. Body gears should have a diametral clearance of 0.002-0.006 in pump body.

When installing the oil pump, vary the number of shims located between pump body and front main bearing cap to provide a backlash of 0.005-0.010 between the crankshaft timing gear and the oil pump drive gear.

Series 300-300U-350-350U-400-400D-W400-W400D-450-450D-W450-W450D

66A. The gear type oil pump, which is gear driven from a pinion on the camshaft, is accessible for removal after removing the engine oil pan. Disassembly and overhaul of the pump is evident after an examination of the unit and reference to Fig. IH620 or IH621. Gaskets between pump cover and body can be varied to obtain the recommended body gear end play. Check the pump parts against the values which follow:
Diametral clearance of gears in pump body:
   Series 300-300U-
      350-350U ....0.005 -0.008
   Series 400-W400-
      450-W450 ....0.004 -0.006
   Series 400D-W400D-
      450D-W450D 0.0060-0.0075
Body gears end play.....0.003 -0.006
Body gear backlash......0.003 -0.006

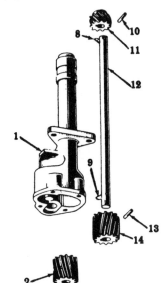

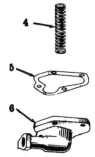

Fig. IH620—Exploded view of oil pump used on non-Diesel engines. Body gear end play is controlled by the number of gaskets (5). On some late models, shaft (12) is carried in renewable bushings.
1. Pump body
2. Follower gear
3. Pressure relief valve
4. Relief valve spring
5. Cover gaskets
6. Pump cover
8. Woodruff key
9. Woodruff key
10. Pin
11. Driver gear
12. Drive shaft
13. Pin
14. Driven gear

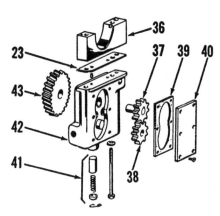

Fig. IH619C—Series 350D and 350DU oil pump is mounted on bottom side of the crankshaft front main bearing cap. Shims (23) control backlash between oil pump gear and crankshaft gear.
36. Main bearing cap, front
37. Driven gear and shaft
38. Idler gear
39. Gasket
40. Cover
41. Oil pressure relief valve
42. Pump body
43. Pump drive gear

OIL PRESSURE RELIEF VALVE

Series 300-300U-350-350U-400-400D-W400-W400D-450-450D-W450-W450D

67. On all models, the spring loaded plunger type oil pressure relief valve is non-adjustable.

On diesel models, the valve is located in the oil filter base as shown in Fig. IH622. On non-diesel models, the valve is contained in the oil pump body as shown in Fig. IH620. Check oil pressure and relief valve against the values which follow:

Oil pressure-psi:

Non-diesels 60-70
Diesels 38-46
Relief valve diameter....0.900-0.901
Relief valve clearance
  in bore 0.004-0.006

Relief valve spring test data:

Non-diesels ....42 lbs. @ $2\frac{3}{32}$ inches
Diesels ....27 lbs. @ $2\frac{3}{32}$ inches

Series 350D-350DU

67A. Oil pressure relief valve and spring are located in oil pump body as shown in Fig. IH619C. Relief valve diameter is 0.6205-0.6215. Relief valve spring should be renewed if oil pressure is insufficient.

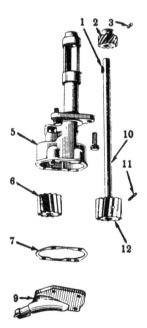

Fig. IH621—Exploded view of the engine oil pump used on series 400D, W400D, 450D and W450D. The oil pressure relief valve is located within the oil filter base as shown in Fig. IH622.

1. Woodruff key
2. Driver gear
3. Pin
5. Pump body
6. Follower gear
7. Gasket
9. Pump cover
10. Drive shaft
11. Pin
12. Driven gear

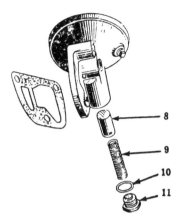

Fig. IH622—Series 400D, W400D, 450D and W450D engine oil pressure relief valve exploded from the oil filter base.

8. Relief valve
9. Spring
10. Gasket
11. Nut

1. Throttle shaft
2. Throttle stop pin
3. Throttle valve screw
4. Throttle valve
5. Throttle shaft bushing
6. Carburetor gasket
7. Throttle body
8. Expansion plug
9. Choke control bracket
10. Bracket clamp
11. Strainer gasket
12. Strainer
13. Expansion plug
14. Throttle shaft bushing
15. Idle outlet plug
16. Venturi
17. Idle needle spring
18. Idle adjusting needle
19. Stop screw spring
20. Throttle stop screw
21. Fuel valve gasket
22. Inlet needle and seat
23. Float axle support
24. Float axle
25. Float
26. Fuel bowl gasket
27. Idle metering jet
28. Main air bleed
29. Discharge jet gasket
30. Main discharge jet
31. Choke lever stop pin
32. Swivel
33. Choke valve shaft
34. Dust seal retainer
35. Dust seal
36. Choke valve
37. Fuel bowl
38. Drip hole filler plug
39. Gasket (distillate or kerosene)
40. Discharge jet nut (distillate or kerosene)
41. Drain cock body (distillate or kerosene)
42. Drain cock assembly (distillate or kerosene)
43. Drain cock stem (distillate or kerosene)
44. Main metering seat (gasoline)
45. Fuel adjustment plug gasket (gasoline)
46. Fuel adjustment plug (gasoline)
47. Needle valve seat (distillate or kerosene)
48. Packing (distillate or kerosene)
49. Needle valve packing nut (distillate or kerosene)
50. Needle valve (distillate or kerosene)
51. Spring
52. Ball

CARBURETOR
(Except L-P Gas)

Series 300-300U-350-350U-400-W400-450-W450

68. Non-diesel models are equipped with an IH 1¼ inch updraft type carburetor. The procedure for disassembling and overhauling the unit is conventional and evident after an examination of Fig. IH623. Refer to an IH parts catalog for calibration data.

69. **ADJUSTMENTS.** Before attempting to adjust the carburetor, first start engine and let it run until thoroughly warmed; then adjust the throttle stop screw (20—Fig. IH623) to obtain a slow idle speed of approximately 425 rpm. Then, turn the idle mixture adjusting screw (18) either way as required to obtain a smooth idle and recheck the slow idle speed. Clockwise rotation of the idle

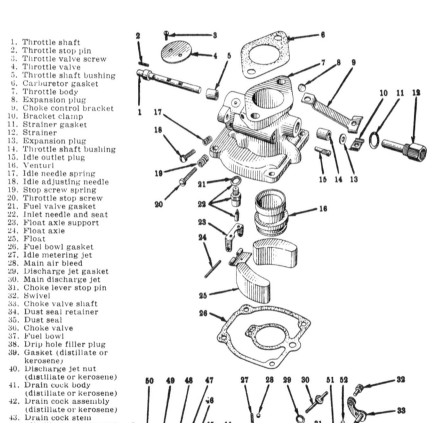

Fig. IH623—Exploded view of the IH 1¼-inch updraft carburetor used on non-Diesel models.

mixture adjusting needle leans the mixture on gasoline models, enriches the mixture on kerosene and distillate burning models.

Clockwise rotation of the main fuel adjustment screw leans the mixture on all models. On gasoline models, the main fuel adjustment screw can be used to reduce the amount of fuel flow when the engine is working under light load conditions; but, when the engine is required to deliver full power, the main adjustment screw must be set five turns open. The adjusting screw seat on gasoline burning models is calibrated to deliver full-power mixture and should not be restricted by the adjustment screw.

On kerosene and distillate burning models the main fuel adjustment screw must be adjusted to control the amount of fuel for best operation under all loaded conditions.

Always recheck the idle mixture after making the main fuel adjustment.

70. **FLOAT SETTING.** Distance from farthest face of float to gasket surface of throttle body when inlet needle is closed is $1\frac{5}{16}$-inch. This float setting provides a fuel level of $\frac{9}{16}$-$2\frac{1}{32}$-inch.

Series 400D-W400D-450D-W450D

71. Diesel models are equipped with an IH model F8 gasoline carburetor which is used for starting the Diesel engine only. Refer to Fig. IH624. When the Diesel engine is in the starting position (running on gasoline) the carburetor float is free and fuel enters the carburetor bowl in the conventional manner via the inlet valve and seat (34). High velocity air entering

8. Tube holder
9. Dust seal retainer
10. Choke valve dust seal
11. Choke valve lever
12. Long choke valve shaft
14. Choke valve
15. Short choke valve shaft
16. Shaft hole plug
17. Screw
18. Air valve
19. Cam
20. Bearing lock
21. Locking shaft bearing
22. Locking shaft dust washer
23. Dust washer retainer
24. Spring
25. Locking shaft
26. Stop pin
27. Air valve spacer
28. Carburetor body
29. Nozzle
30. Metering well
31. Fuel inlet tube
32. Gasket
33. Float pivot
34. Inlet needle and seat
35. Seal gasket
36. Washer
37. Screw
42. Drain valve
43. Strainer
44. Strainer gasket
45. Fuel bowl
47. Float lever
48. Float
49. Lower leaf spring
50. Upper leaf spring
51. Float lever spacer
52. Pivot screw gasket

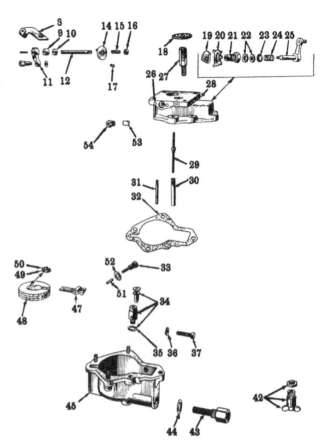

Fig. IH624—Exploded view of model F8 carburetor used for starting series 400D-W400D-450D-W450D.

around choke valve (14) passes around air valve (18) which is open at all times. The vacuum around air valve (18) draws fuel from nozzle (29). When the engine is switched from the gasoline starting position to Diesel operation, the inlet needle valve is closed by cam action and the carbure-

tor ceases to function.

The space between the air valve and the carburetor body should be 3/32-7/64-inch. Obtain the recommended fuel level of 13/32-7/16-inch below the bowl rim by bending the float tang or by varying the number of gaskets under the inlet needle cage.

LP-GAS SYSTEM

International Harvester tractors are available with factory installed LP-Gas systems using Ensign equipment. The fuel tank should never be completely filled; there should always be 10-20% of the tank capacity allowed for expansion due to a possible temperature rise.

It is important when starting LP-Gas tractors to open the vapor valve on the supply tank SLOWLY; if opened too fast, the fuel supply to the regulator will be shut off. Too rapid opening of vapor or liquid valves may cause freezing.

SYSTEM ADJUSTMENTS

72. The LP-Gas carburetor and regulator have three points of mixture ad-

Fig. IH625—Side view of LP-Gas engine showing the carburetor and regulator installation. Note the location of adjustments.

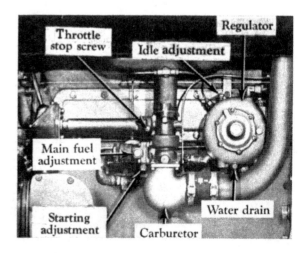

justment plus a throttle stop screw. Refer to Fig. IH625. Initial starting adjustment setting is 1⅓ turns open for series 300, 300U, 350 and 350U, ⅔-turn open for series 400, W400, 450 and W450. This adjustment, however, can be varied slightly, if satisfactory cold starts are not obtained. Initial main fuel adjustment setting is 3¼ turns open for series 300, 300U, 350 and 350U, 3¾ turns open for series 400, W400, 450 and W450. Initial idle adjustment setting is one turn open for all models. Start engine, allow to warm up and proceed as follows:

THROTTLE STOP SCREW. Stop screw on carburetor throttle should be adjusted to provide an engine low idle speed of 400-450 rpm for series 300 and 300U, 475-525 rpm for series 350, 400, W400, 450 and W450, 500-550 rpm for series 350U.

IDLE ADJUSTMENT. With the choke open, engine warm and the throttle stop screw set, rotate the idle adjusting screw until the engine speed is highest and operation is smoothest.

MAIN FUEL ADJUSTMENT. With the engine running at high idle rpm, turn the main fuel adjustment either way as required to obtain the smoothest operation. When tractor is operating under load, it may be necessary to vary the adjustment slightly to obtain the smoothest operation.

LP-GAS FILTER

73. Filters used in LP-Gas systems should be able to stand high pressure without leakage. Refer to Fig. IH626. When major engine work is being performed, it is advisable to remove the lower part of the filter, thoroughly clean the interior and renew the treated paper filter cartridge.

LP-GAS REGULATOR

74. **HOW IT OPERATES.** Fuel from the supply tank enters the regulating unit inlet (A—Fig. IH627) at tank pressure and is reduced to about 4 psi at the high pressure reducing valve

(C) after passing through the strainer (B). Flow through high pressure reducing valve is controlled by the adjacent spring and diaphragm. When the liquid fuel enters the vaporizing chamber (D) via the valve (C), it expands rapidly and is converted from a liquid to gas by heat from the water jacket (E) which is connected to the cooling system of the engine. The vaporized gas then passes at a pressure slightly below atmospheric via the low pressure reducing valve (F) into the low pressure chamber (G) where it is drawn off to the carburetor via outlet (H). The low pressure reducing valve is controlled by the larger diaphragm and small spring.

Fuel for the idling range of the engine is supplied from a separate outlet which is connected by tubing to a separate idle fuel connection on the carburetor.

TROUBLE SHOOTING

75. **SYMPTOM.** Engine will not idle with Idle Mixture Adjustment Screw in any position.

CAUSE AND CORRECTION. A leaking valve or gasket is the cause of the trouble. Look for leaking low pressure valve caused by deposits on valve or seat. To correct the trouble, wash the valve and seat in gasoline or other petroleum solvent.

If foregoing remedy does not correct the trouble, check for leak at high pressure valve by connecting a low reading (0-20 psi) pressure gauge to plug opening on front of regulator. If the pressure increases after a warm engine is stopped, it proves a leak in

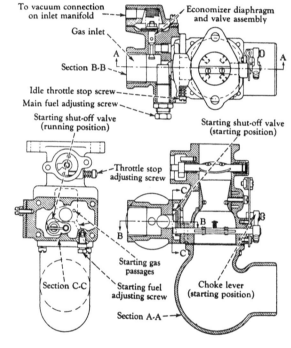

Fig. IH626A—Sectional views of LP-Gas carburetor. The procedure for overhauling the unit is conventional.

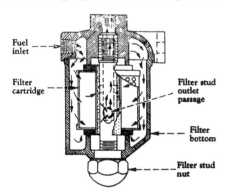

Fig. IH626—Sectional view of LP-Gas fuel filter. The unit contains a renewable treated paper element.

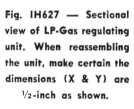
Fig. IH627 — Sectional view of LP-Gas regulating unit. When reassembling the unit, make certain the dimensions (X & Y) are ½-inch as shown.

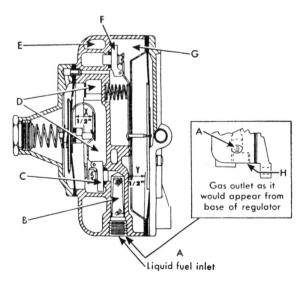

the high pressure valve. Normal pressure is 3½-5 psi.

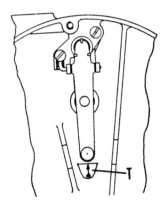

Fig. IH628—Location of post or boss with stamped arrow for the purpose of setting the fuel inlet valve lever.

76. **SYMPTOM.** Cold regulator shows moisture and frost after standing.

CAUSE AND CORRECTION. Trouble is due either to leaking valves as per paragraph 74 or the valve levers are not properly set. For information on setting of valve lever refer to paragraph 78.

77. **SYMPTOM.** Frost on filter.

CAUSE AND CORRECTION. Filter element is clogged and must be renewed.

REGULATOR OVERHAUL

78. If an approved station is not available, the regulator can be overhauled as follows:

Remove the unit from the engine and completely disassemble. Thoroughly wash all parts and blow out

all passages with compressed air. Inspect each part carefully and discard any that are worn.

Before reassembling the unit, note dimension (X—Fig. IH627) which is measured from the face on the high pressure side of the casting to the inside of the groove in the valve lever when valve is held firmly shut. If dimension (X) which can be measured with Ensign gauge No. 8276 or with a depth rule is more or less than ½-inch, bend the lever until this setting is obtained. A boss or post (T—Fig. IH628) is machined and marked with an arrow to assist in setting the lever. Be sure to center the lever on the arrow before tightening the screws holding the valve block. The top of the lever should be flush with the top of the boss or post (T).

DIESEL FUEL SYSTEM

The Diesel fuel system consists of three basic components; the fuel filters, injection pump and injection nozzles. When servicing any unit associated with the fuel system, the maintenance of absolute cleanliness is of utmost importance. Of equal importance is the avoidance of nicks or burrs on any of the working parts.

Probably the most important precaution that service personnel can import to owners of Diesel powered tractors, is to urge them to use an approved fuel that is absolutely clean and free from foreign material. Extra precaution should be taken to make certain that no water enters the fuel storage tanks. This last precaution is based on the fact that all Diesel fuels contain some sulphur. When water is mixed with sulphur, sulphuric acid is formed and the acid will quickly

erode the closely fitting parts of the injection pump and nozzles.

NOTE: Series 400D, W400D, 450D and W450D are equipped with injection nozzles and injection pump manufactured by International Harvester; whereas, on series 350D and 350DU, the injection pump is manufactured by Roosa-Master and the injectors by C.A.V.

TROUBLE SHOOTING
Series 400D-W400D-450D-W450D

96. The unit construction of the Diesel system components permits removal of faulty units and installation of new (or reconditioned) units without disturbing the calibration of others. However, before removing a suspected faulty unit, it is advisable to

make a systematic check of the complete system to make certain which unit (or units) are at fault. To make such a check, use either a Kiene, Bacharach or Buda test kit and closely follow the instructions with the kit.

QUICK CHECKS
All Models

96A. If the engine does not run properly and the fuel system is suspected as the source of trouble, refer to the accompanying trouble shooting chart and locate points which require further checking. Many of the chart items are self-explanatory; however, if the difficulty points to the fuel filters, injection nozzles and/or injection pump

DIESEL SYSTEM TROUBLE-SHOOTING CHART

	Lack of Fuel	Engine Surging or Rough	Cylinders Uneven	Engine Smokes or Knocks	Injection Pump Does Not Shut Off	Engine Dies at Low Speed	Loss of Power
Defective Speed Control Linkage	★				★		★
Air in Fuel System	★			★		★	
Clogged Filters and/or Water Trap	★					★	★
Fuel Lines Leaking or Clogged	★	★	★			★	★
Friction in Injection Pump		★					★
Inferior or Contaminated Fuel		★		★		★	★
Faulty Injection Pump Timing		★		★		★	★
Defective Nozzle or Injector		★	★	★			★
Faulty Governor and/or Linkage Adjustment		★		★	★		★
Sticking Plunger		★				★	★
Faulty Primary Pump and/or Pump Gaskets	★						★
Faulty Distribution of Fuel	★	★	★	★			★
Injection Pump Not Turning	★						
Friction in Governor		★					

refer to the appropriate sections which follow:

FILTERS AND BLEEDING
Series 400D-W400D-450D-W450D

97. As shown in Figs. IH635, 636 and 637, the fuel filtering system consists of a water trap and two stages of renewable element type filters.

98. **MAINTENANCE.** Start engine and note position of hand on the fuel pressure gage. If hand is in the "RED" area, remove the water trap and thoroughly clean same. Reinstall the water trap, bleed the system as outlined in paragraph 99 and recheck the pressure gage reading. If the gage hand is still in the "RED", renew the auxiliary fuel filtering element and clean the primary pump filtering screen. To renew the element, remove the filter case and element and thoroughly clean the case and filter base. Install plates (4 & 6—Fig. IH636) on new element making certain that the plates slide into wire coil inside of the element and that face of top plate is installed with word "TOP" toward top of tractor. Remove the primary pump filter screen as shown in Fig. IH638 and thoroughly clean same using kerosene or clean Diesel fuel. Reassemble the parts, bleed the system and recheck the pressure gage reading. If gage hand is still in the "RED", it will be necessary to renew the final fuel filtering element. The procedure for renewing the final filtering element is evident after an examination of the unit and reference to Fig. IH636.

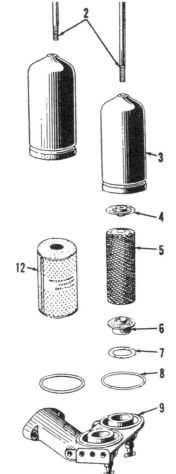

Fig. IH636—Exploded view of IH fuel filters used on series 400D, W400D, 450D and W450D. Case studs (2) are not interchangeable.

2. Case studs	7. Gasket
5. Auxiliary filter element	8. Gasket
6. Filter element bottom plate	9. Filter base
	12. Final filter element

CAUTION: The filter case studs (2) are not interchangeable and must be installed in their original position.

If after renewing the final fuel filtering element, the pressure gage hand is still in the "RED", it will be necessary to renew or recondition the primary pump as outlined in paragraph 107.

99. **BLEEDING.** The fuel system should be purged of air whenever the fuel filters have been removed or when the fuel lines have been disconnected. To bleed the system, first make certain that fuel level in tank is above the auxiliary fuel filter, open the fuel tank shut-off valve and proceed as follows: Open the water trap vent (1—Fig. IH635) and the auxiliary fuel filter vent cock (2); close the vents when the fuel flows free of air. Start the engine and run on the gasoline cycle. Open the final fuel filter vent cock (3); close the vent when fuel flows free of air. Advance the engine speed control lever slightly but do not touch the compression release lever. Open each nozzle vent (4) individually; close the vent when the fuel flows free of air.

Series 350D-350DU

99A. The fuel filtering system consists of an "edge-type" primary filter and water trap, a renewable element type secondary filter, and a sealed type final filter.

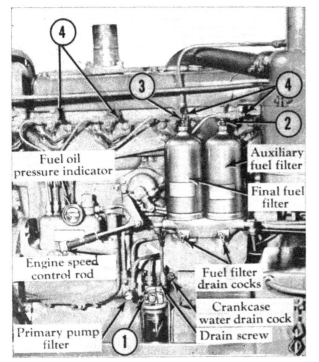

Fig. IH635—Series 400D, W400D, 450D and W450D fuel system consists of three basic units; the injection nozzles, injection pump and fuel filters.

1. Water trap vent
2. Auxiliary filter vent
3. Final filter vent
4. Nozzle vent

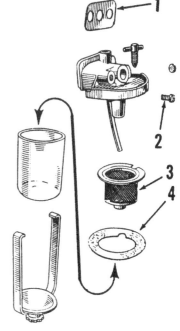

Fig. IH637—Exploded view of IH water trap used on series 400D, W400D, 450D and W450D. It is necessary to bleed the fuel system after cleaning the water trap.

1. Gasket	3. Screen
2. Vent screw	4. Gasket

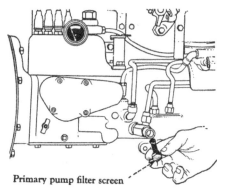

Fig. IH638—Removing series 400D, W400D, 450D and W450D primary pump filter screen. Zero or low fuel pressure can be caused by a clogged screen.

99B. MAINTENANCE. With the fuel shut-off valve closed (Fig. IH638A), remove drain plug (1—Fig. IH638B) and drain off any accumulation of water and dirt. This should be done daily before tractor is operated. After every 60 hours of operation or more often in severe dust conditions, the primary filter and water trap (PF & WT) should be disassembled and thoroughly cleaned in clean Diesel fuel or a suitable solvent. When reassembling, use a new gasket and be sure all connections are tight. Bleed the fuel system as in paragraph 99C.

Every 250 hours of operation or more often in severe dust conditions, remove the element from secondary filter (SF) and thoroughly clean the filter case with clean Diesel fuel or a suitable solvent. Install a new filter cartridge and bleed the fuel system as in paragraph 99C.

Every 500 hours of operation or more often in severe dust conditions, renew the complete final fuel filter (FF—Fig. IH638C) and bleed the fuel system as outlined in paragraph 99C.

99C. BLEEDING. When the fuel flow has been interrupted or when fuel lines have been disconnected, bleed air from the system as follows:

Open the fuel tank shut-off valve (Fig. IH638A) and remove the vent plug (2—Fig. IH638B) from top of the secondary fuel filter. When the fuel flows in a solid stream without air bubbles, reinstall the vent plug.

Remove vent plug (3—Fig. IH638C) from top of the final fuel filter. When the fuel flows in a solid stream without air bubbles, reinstall the vent plug.

Loosen the pipe plug (4) on the injection pump, crank engine with the starting motor until the fuel flows in a solid stream without air bubbles, then tighten the pipe plug.

Loosen the high pressure fuel line connections at the injectors and crank engine with the starting motor until fuel appears. Tighten the fuel line connections and start engine.

I. H. INJECTION NOZZLES
Series 400D-W400D-450D-W450D

Early production series 400D and W400D tractors were factory equipped with orifice type injection nozzles shown exploded in Fig. IH639. Beginning with engine serial number D-264-21481, the 400D and W400D tractors were factory equipped with open type injection nozzles shown exploded in Fig. IH640. All series 450D and W450D tractors are equipped with open type injection nozzles.

IH nozzle field packages (No. 271 745 R91) are available for converting the old orifice type nozzles to the new open type. CAUTION: Open type injection nozzle valves must not be mixed with the orifice type in any one engine; to do so would result in extremely rough engine operation. Therefore, when making the conversion, it will be necessary to obtain and install four of the 271 745 R91 packages. Also, when making this conversion, it is necessary to enlarge the exit opening in the old nozzle to 0.450-0.455. A

Fig. IH638B—Right side view of series 350D engine, showing the installation of the primary filter and water trap (PF & WT), secondary filter (SF) and the engine oil filter. Series 350DU is similar.

Fig. IH638A—Fuel tank shut-off valve location on series 350D. On series 350DU, the valve is similarly located at bottom of fuel tank.

Fig. IH638C—Left side view of series 350D engine, showing the installation of the final fuel filter (FF). The filter location on series 350DU is similar.

diagram of this modification is included in each field package.

After converting from the orifice type to the open type nozzles, it may be necessary to readjust the injection pump governor. Refer to paragraph 110.

100. **TESTING-WARNING:** Fuel leaves the injection nozzles with sufficient force to penetrate the skin. When testing, keep your person clear of the nozzle spray.

If the engine does not run properly, or not at all, and the checks outlined in paragraph 96A point to a faulty injection nozzle, test the nozzle as follows: Start engine and run on Diesel cycle. Open and close the nozzle bleeder valves (4—Fig. IH635) one at a time until the cylinder (or cylinders) are found that least affect engine performance. If a cylinder is misfiring,

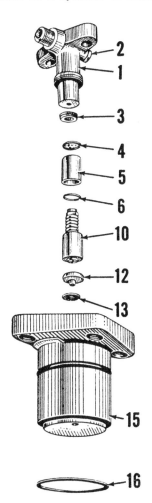

Fig. IH639—Exploded view of the old style orifice type IH injection nozzle used in production on series 400D and W400D prior to engine serial No. 21481. The nozzle valve (10) can be overhauled providing suitable nozzle testing equipment is available.

1. Nozzle fitting	10. Nozzle valve
2. Bleeder valve	12. Injector plate
3. Dust seal	13. Gasket
4. Screen	15. Nozzle body
5. Spacer	16. Gasket
6. Gasket	

it is reasonable to suspect a faulty nozzle; however, a similar condition can be caused by a defective valve in the distributor block.

If nozzle appears to be faulty, remove the nozzle as in paragraph 101, place nozzle in a test stand and check the nozzle against the following specifications:

Minimum opening pressure....700 psi
No leakage below............400 psi

If nozzle requires overhauling, refer to paragraph 102 or 103.

101. **REMOVE AND REINSTALL.** To remove any injection nozzle, first remove dirt from nozzle, injection pump and cylinder head; then, disconnect and remove the injector pipe. Cover all openings with tape or composition caps to prevent the entrance of dirt or other foreign materials. Remove the two nozzle retaining stud nuts, lift nozzle from cylinder head and remove the nozzle body dust seal. IH service tool No. 1 020 284 R91 is available for pulling a stuck nozzle.

102. **OVERHAUL (ORIFICE TYPE).** Remove cap screws retaining the nozzle fitting to the nozzle body and remove the fitting (1—Fig. IH639). Remove the dust seal (3), screen (4), spacer (5) and gaskets. Remove the nozzle valve and spring unit (10). Remove injector plate (12) and gasket (13). Thoroughly clean and inspect all parts, and renew any which are damaged. Inspect the orifice size in the injection plate (12). The orifice size, which should be 0.036-0.037, can be checked with a piano wire of the specified size. Always renew a questionable orifice plate. The nozzle opening pressure is controlled by the adjustment of the nozzle valve and spring. This adjustment should not be disturbed unless nozzle testing equipment is available. If such equipment is not available and the valve assembly is suspected as the source of trouble, install a new nozzle valve which is available as a pre-adjusted unit. If nozzle testing equipment is available, proceed as follows:

Disassemble the nozzle valve (10) and thoroughly clean parts in clean Diesel fuel. Inspect face and seat of nozzle valve with a magnifying glass. If face and seat are not excessively worn, they can be reconditioned by lapping by using carborundum H-400 fine until polished surfaces are obtained. Check the nozzle valve spring against the following values:

Free length 0.802-0.834
Test length 5/8-inch
Test load (lbs.).......33.06-36.54

Renew the nozzle valve spring if it is rusted, discolored or if it does not meet the foregoing pressure specifications.

Reassemble the nozzle valve and nozzle and place nozzle in a test fixture. Check nozzle against the following values:

Opening pressure
(new spring) ....740-750 psi

Opening pressure
(used spring)........700-710 psi

No leakage below........ 400 psi

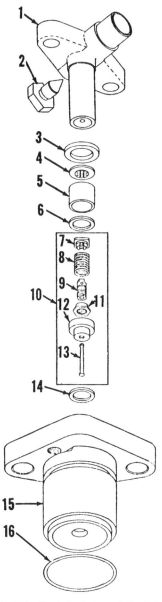

Fig. IH640—Exploded view of the late production open type IH injection nozzle. Nozzle opening pressure is non-adjustable.

1. Nozzle fitting	9. Stop screw
2. Bleeder valve	10. Valve assembly
3. Dust seal	11. Spring lower seat
4. Screen	12. Valve seat
5. Sleeve	13. Valve
6. Sleeve gasket	14. Valve gasket
7. Valve upper seat	15. Nozzle body
8. Nozzle spring	16. Gasket

If opening pressure is too low, disassemble the nozzle and tighten the nozzle valve adjusting nut; if pressure is too high, loosen the nut. If leakage occurs below 400 psi, the nozzle valve and seat requires renewal or further lapping.

When reassembling, tighten the nozzle fitting cap screw to a torque of 30-35 ft.-lbs. When installing the nozzle reverse the removal procedure and tighten the retaining stud nuts to a torque of 20-25 ft.-lbs.

103. OVERHAUL (OPEN TYPE). Remove cap screws retaining the nozzle fitting (1 — Fig. IH640) to nozzle body and remove the fitting. Invert the nozzle body and remove the dust seal (3), screen (4), spacer (5) and gaskets. Remove the nozzle valve assembly (10) and gasket (14). Thoroughly clean and inspect all parts and renew any which are damaged. The nozzle opening pressure is non-adjustable, but the valve lift is adjustable. Normally, the valve should not be disassembled; a pre-assembled unit is available for service installation. If however, the valve is to be disassembled, proceed as follows: Unlock and screw the lower spring seat (11) upward and away from valve seat (12) a few turns. Then turn the upper spring seat (7) clockwise until the lower spring seat (11) is again against the valve seat (12). Then compress spring (8) and remove upper spring seat (7), spring (8) and valve (13). Remove stop screw (9) and seat (11) from valve seat (12). Examine all parts and renew any which are damaged. Seat (12) and valve (13) are not available separately. Reassemble the valve unit.

NOTE: Beginning with engine serial No. 21481, series 400D and W400D as well as all series 450D and W450D tractors are fac-

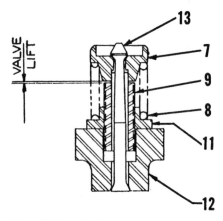

Fig. IH641—Sectional view of the open type nozzle valve. Nozzle opening pressure is non-adjustable, but valve lift should be adjusted as outlined in paragraph 103A or 103B. Refer to Fig. IH640 for legend.

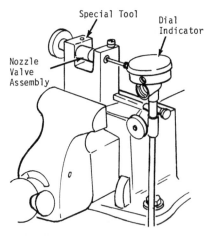

Fig. IH642—Using dial indicator and IH special tool No. 1020 334 R91 to check the lift of the open type nozzles.

tory fitted with the open type nozzles and the valve lift should be adjusted as in the following paragraph 103A. On early series 400D and W400D where the original orifice type nozzles have been converted to the open type, refer to paragraph 103B.

103A. To adjust the valve lift on series 450D & W450D as well as series 400D and W400D beginning with serial No. 21481, proceed as follows: Clamp the valve seat (12) lightly in a soft jawed vise and turn the spring lower seat (11) counter-clockwise, slightly away from valve seat (12). Refer to Fig. IH641. Turn upper spring seat (7) counter-clockwise until there is no valve lift. When this point of no valve lift is reached, turn the upper spring seat (7) clockwise ⅜-turn and lock in that position by tightening the lower spring seat against the valve seat. Check the actual valve lift which should be 0.012-0.013 as shown. This lift can be checked with a dial indicator in a manner similar to that shown in Fig. IH642. If the valve lift is not as specified, loosen the lower spring seat, turn the upper seat a slight amount as required and recheck.

103B. On series 400D and W400D prior to engine serial No. 21481, wherein the original production orifice type nozzles have been converted to the open type using field packags No. 271 745 R91, the procedure for adjusting the valve lift is the same as that outlined in paragraph 103A, but the amount of lift should be adjusted to 0.007-0.008 **NOT** 0.012-0.013. This reduced valve lift is necessary to compensate for the injection pump and governor setting used with the original orifice type nozzles.

103C. Assemble all open type nozzles by reversing the disassembly procedure and be sure sleeve (5—Fig. IH 640), gasket (6) and valve assembly (10) are all the way down in the nozzle body before tightening the nozzle fitting. Failure to do so may result in breaking the spring upper seat (7). Tighten the nozzle fitting cap screws to a torque of 30-35 ft.-lbs.

Place the assembled nozzle in a test fixture and check the opening pressure as follows: When nozzle is equipped with a new spring, the opening pressure will be 900-1000 psi. The opening pressure is not adjustable; however, the nozzle will continue to function satisfactorily so long as the opening pressure does not fall below 700 psi. To insure smooth engine operation, the opening pressure in any set of nozzles for one engine should not vary more than 100 psi.

Install the nozzle by reversing the removal procedure and tighten the retaining stud nuts to a torque of 20-25 ft.-lbs.

C.A.V. INJECTION NOZZLES
Series 350D-350DU

The D193 engines are factory fitted with C.A.V. BKB 70S 640 injector assemblies which include delay type BDN 12S D12 nozzle units shown in Fig. IH642A.

WARNING: Fuel leaves the injection nozzles with sufficient pressure (1750-1850 psi) to penetrate the skin. When testing, keep your person clear of the nozzle spray.

103D. **TESTING AND LOCATING A FAULTY NOZZLE.** If the engine does not run properly, or not at all, and the quick checks outlined in paragraph 96A point to a faulty injector, locate the faulty unit as follows:

If one engine cylinder is misfiring, it is reasonable to suspect a faulty injector. Generally, a faulty injector can be located by loosening the high pressure line fitting on each nozzle holder in turn, thereby allowing fuel to escape at the union rather than enter the cylinder. As in checking

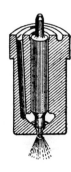

Fig. IH642A — Sectional view of C. A. V. injection nozzle used on D-193 engines. Injectors are adjusted to provide a normal cracking pressure of 1750-1850 psi.

spark plugs in a spark ignition engine, the faulty unit is the one which, when its line is loosened, least affects the running of the engine.

103E. Remove the suspected injector from the engine as outlined in paragraph 103K. If a suitable nozzle tester is available, check the unit as in paragraphs 103F, 103G, 103H and 103J. If a tester is not available, reconnect the fuel line to the injector and with the nozzle tip directed where it will do no harm, crank the engine with the starting motor and observe the nozzle spray pattern.

If the spray pattern is ragged, unduly wet, streaky and/or not symetrical or, if nozzle dribbles, the nozzle valve is not seating properly. Send the complete nozzle and holder assembly to an official diesel service station for overhaul.

103F. **NOZZLE TESTER.** A complete job of testing and adjusting the injector requires the use of a special tester such as that shown in Fig. IH-642B. The nozzle should be tested for opening pressure, seat leakage and spray pattern. Operate the tester lever until oil flows and attach the nozzle and holder assembly.

Note: Only clean, approved testing oil should be used in the tester tank.

Close the tester valve and apply a few quick strokes to the lever. If undue pressure is required to operate the lever, the nozzle valve is plugged and same should be serviced as in paragraph 103L.

103G. Opening Pressure. While operating the tester handle, observe the gage pressure at which the spray occurs. The gage pressure should be 1750-1850 psi. If the pressure is not

as specified, remove the nozzle protecting cap, exposing the pressure adjusting screw and locknut. Loosen the locknut and turn the adjusting screw as shown in Fig. IH642B either way as required to obtain an opening pressure of 1750-1850 psi. Note: If a new pressure spring has been installed in the nozzle holder, adjust the opening pressure to 1900-2000 psi. Tighten the locknut and install the protecting cap when adjustment is complete.

103H. Seat Leakage. The nozzle valve should not leak at a pressure less than 1600 psi. To check for leakage, actuate the tester handle slowly and as the gage needle approaches 1600 psi, observe the nozzle tip for drops of fuel. If drops of fuel collect at pressures less than 1600 psi the nozzle valve is not seating properly and same should be serviced as in paragraph 103L.

103J. Spray Pattern. Operate the tester handle at approximately 100 strokes per minute and observe the nozzle spray pattern. If the spray pattern is unduly wet, streaky and/or ragged, the nozzle valve should be serviced as in paragraph 103L.

103K. **REMOVE AND REINSTALL.** Before loosening any fuel lines, wash the nozzle holder and connections with clean diesel fuel or kerosene. After disconnecting the high pressure and leak-off lines, cover open ends of connections with composition caps or tape to prevent the entrance of dirt or other foreign material. Remove the nozzle holder screws and carefully withdraw the nozzle from cylinder head, being careful not to strike the tip end of the nozzle against any hard surface.

Thoroughly clean the nozzle recess in the cylinder head before reinserting the nozzle and holder assembly. It is important that the seating surfaces of recess be free of even the smallest particle of carbon which could cause the unit to be cocked and result in blowby of hot gases. No hard or sharp tools should be used for cleaning. A piece of wood dowel or brass stock properly shaped is very effective. Do not reuse the copper ring gasket (12—Fig. IH642C), always install a new one. Tighten the nozzle holder screws to a torque of 14-16 Ft.-Lbs.

103L. **MINOR OVERHAUL (CLEANING) OF NOZZLE VALVE AND BODY.** Hard or sharp tools, emery cloth, crocus cloth, grinding compounds or abrasives of any kind should NEVER be used in the cleaning of nozzles. A nozzle cleaning and main-

tenance kit is available through any C. A. V. Service Agency under the number of ET.140.

Wipe all dirt and loose carbon from the nozzle and holder assembly with a clean, lint free cloth. Carefully clamp nozzle holder assembly in a soft jawed vise and remove the protecting cap (13—Fig. IH642C). Loosen jam nut (1) and back-off the adjusting screw (3) enough to relieve load from spring (5). Remove the nozzle cap nut (7) and nozzle body (9). Normally, the nozzle valve (8) can be easily withdrawn from the nozzle body. If the valve cannot be easily withdrawn, soak the assembly in fuel oil, acetone, carbon tetrachloride or similar carbon solvent to facilitate removal. Be careful not to permit the valve or body to come in contact with any hard surface.

Examine the nozzle body and remove any carbon deposits from ex-

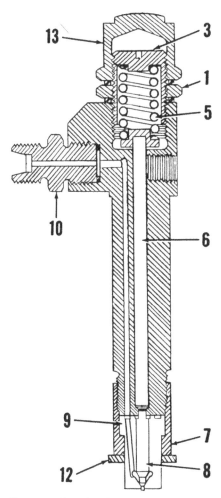

Fig. IH642C — Sectional view of a typical C. A. V. injector. When installing the injector in the cylinder head, be sure to use a new copper washer (12).

1. Jam nut	8. Nozzle valve
3. Adjusting screw	9. Nozzle body
5. Spring	10. Fuel inlet connection
6. Valve spindle	12. Special copper washer
7. Nozzle cap nut	13. Protecting cap

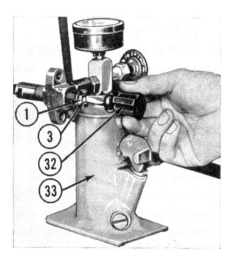

Fig. IH642B—Adjusting the nozzle opening pressure, using a suitable nozzle tester. Normal opening pressure is 1750-1850 psi.

1. Nut	32. Screw driver
3. Adjusting screw	33. Nozzle tester

terior surfaces using a brass wire brush (C. A. V. No. ET.068). The nozzle body must be in good condition and not blued due to overheating. All polished surfaces should be relatively bright, without scratches or dull patches. Pressure surfaces (A, B & J—Fig. IH642D) must be absolutely clean and free from nicks, scratches or foreign material, as these surfaces must register together to form a high pressure joint.

Clean out the small fuel feed channels (C), using a small diameter wire. Insert the special groove scraper (C. A. V. No. ET.071) into nozzle body until nose of scraper locates in fuel gallery (F); then press nose of scraper hard against side of cavity and rotate scraper to clean all carbon deposits from the gallery. Clean all carbon from valve seat (G), using seat scraper (C. A. V. No. ET.070).

Using pintle hole cleaner (C. A. V. No. ET.069) and appropriate size probe, pass the probe down the bore of the nozzle body until probe protrudes through the orifice; then rotate the probe until all carbon is cleared. Examine the pintle and seat end of the nozzle valve and remove any carbon deposits using a brass wire brush

(C. A. V. No. ET. 068). Use extreme care, however, as any burr or small scratch may cause valve leakage or spray pattern distortion. If valve seat (M—Fig. IH642D) has a dull circumferential ring indicating wear or pitting or if valve is blued, the valve and body should be turned over to an official diesel service station for possible overhaul.

Before reassembling, thoroughly rinse all parts in clean diesel fuel and make certain that all carbon is removed from the nozzle holder nut. Install nozzle body and holder nut, making certain that the valve stem is located in the hole of the holder body. Tighten the holder nut.

Note: Over-tightening may cause distortion and subsequent seizure of the nozzle valve.

Test the injector as in paragraphs 103F, 103G, 103H and 103J. If the nozzle does not leak under 1600 psi, and if the spray pattern is satisfactory, the nozzle is ready for use. If the nozzle will not pass the leakage and spray pattern tests, renew the nozzle valve and seat, which are available only in a matched set; or, send the nozzle and holder assembly to an official diesel service station for a complete overhaul which includes reseating the nozzle valve cone and seat.

103M. **OVERHAUL OF NOZZLE HOLDER.** (Refer to Fig. IH642C). Remove cap (13). Remove jam nut (1) and adjusting screw (3). Withdraw spring (5) and spindle (6). Thoroughly wash all parts in clean diesel fuel and examine the end of the spindle which contacts the nozzle valve stem for any irregularities. If the contact surface is pitted or rough, renew the spindle. Renew any other questionable parts.

Reassemble the nozzle holder and leave the adjusting screw locknut loose until after the nozzle opening pressure has been adjusted as outlined in paragraph 103G.

ENERGY CELLS
Series 350D-350DU

103N. **R&R AND CLEAN.** The necessity for cleaning the energy cells is usually indicated by excessive exhaust smoking, or when the fuel economy drops. As shown in Fig. IH642H, any energy cell can be removed without removing manifold or filters.

An emergency job of cleaning the cells (Fig. IH642J) can be accomplished by removing the cell cap (99) and using a hooked wire to form a scraper.

To remove the complete energy cell, first remove the threaded plug (97) and cap (98). Using a pair of thin nosed pliers, grip the tip of the energy cell cap (99) and pull cap out of chamber. Cell (100) can be removed by screwing a puller bolt into cell and pulling cell from chamber. The energy cell can also be bumped out by first removing the injector; then, bumping cell out with a brass drift inserted through the injection nozzle opening.

The removed parts can be cleaned in an approved carbon solvent. After parts are cleaned, visually inspect them for cracks and other damage. Renew any damaged parts. Inspect the mating surfaces of the cell body and the cell cap for being rough and pitted. The surfaces can be reconditioned by lapping with valve grinding compound. Make certain that the energy cell seating surface in cylinder head is clean and free from carbon deposits.

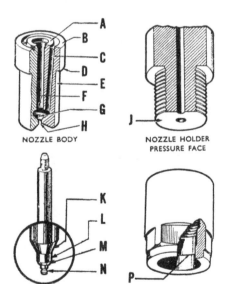

Fig. IH642D—Disassembled views of nozzle and holder, showing various points for detailed inspection and thorough cleaning.

A. Nozzle body pressure face
B. Nozzle body pressure face
C. Fuel feed hole
D. Shoulder
E. Nozzle trunk
F. Fuel gallery
G. Valve seat
H. Pintle orifice
J. Holder pressure face
K. Valve cone
L. Stem
M. Valve seat
N. Pintle
P. Nozzle retaining shoulder

Fig. IH642H—Side view of series 350D and 350DU engine showing the installation of the energy cells. A sectional view of the energy cell installation is shown in Fig. IH642J.

When installing the energy cell, tighten the threaded plug enough to insure an air tight seal.

PRECOMBUSTION CHAMBERS

Series 400D-W400D-450D-W450D

104. **REMOVE AND REINSTALL.** The precombustion chambers can be pulled from cylinder head after removing the respective nozzle assembly. Pre-cup Puller No. 1 020 310 R91 will facilitate removal of the precombustion chamber.

When installing the precombustion chamber, make certain that side stamped "TOP" is installed toward top of engine.

I.H. INJECTION PUMP

Series 400D-W400D-450D-W450D

The subsequent paragraphs will outline ONLY the injection pump service work which can be accomplished without the use of special, costly pump testing equipment. If additional service work is required, the pump should be turned over to an official Diesel service station for overhaul. Inexperienced service personnel should never attempt to overhaul a Diesel injection pump.

The unit construction of the pump permits removal of faulty components and installation of new (or reconditioned) units without disturbing the calibration of others.

If the quick checks outlined in paragraph 96A point to a faulty injection pump, refer to the appropriate following paragraphs:

105. **TIMING.** For normal operation, injection timing is satisfactory when pointer (D—Fig. IH643) is in register with the zero graduation on the pump drive gear as shown. It may, however, be necessary to change this timing slightly to obtain the desired running condition for various load requirements and different grades of fuel. For any given load condition and grade of fuel, the optimum timing is when engine speed is maximum with a clean exhaust.

Note: The following timing procedure is outlined on the assumption that the injection pump drive gear is properly timed with respect to the timing gear train idler gear. Refer to paragraph 50.

To check and adjust the injection pump timing when pump is installed, first crank engine until No. 1 cylinder is coming up on compression stroke and notch "DC" on crankshaft pulley is in register with the pointer on the crankcase front cover. Remove the injection pump drive gear cover from front of timing gear case. The pump is properly timed to the pump drive gear if the chisel marked gear hub groove (A—Fig. IH643) is in line (or within 20 degrees either way) with the zero degree graduation on the injection pump drive gear as shown. If the timing marks are not as specified, it will be necessary to re-position the drive gear hub, with respect to the drive gear, as follows: Remove the three cap screws (B) and turn the drive gear hub (C) until chisel marked groove in the hub is in the position outlined above. Position the timing indicator (D) so that pointer is in register with the zero degree graduation on the gear and install the cap screws (B).

Start the engine and check the engine speed. Stop the engine and shift the timing pointer either way and recheck the engine speed. Continue this operation until maximum engine speed is obtained for any given load, and engine operation is smooth with a clean exhaust.

106. **INJECTION PUMP-R&R.** To remove the complete injection pump unit, first shut off the fuel supply and thoroughly clean dirt from pump, fuel lines and connections. Disconnect the fuel lines, nozzle pipes and speed control rod from pump, and remove glass bowl from water trap. Remove the pump gear cover from timing gear case and remove the timing pointer by taking out the three cap screws (B—Fig. IH643) which hold the pointer, gear and hub together. Remove the cap screws which retain pump to crankcase front cover and remove the pump.

Note: Three of the pump retaining cap screws are located inside the timing gear housing and must be removed through the openings in the pump drive gear.

To reinstall the injection pump, crank engine until No. 1 piston is coming up on compression stroke and the notch marked "DC" on the crankshaft pulley is in register with pointer on the crankcase front cover. Install pump. Turn the injection pump gear hub (C) so that the chisel marked groove (A) on the gear hub is in line with the zero degree graduation on the drive gear as shown. Install the timing indicator with pointer at the 0° position and bolt the gear hub, gear and indicator together in this position.

Recheck the injection timing as outlined in paragraph 105.

107. **PRIMARY PUMP-R&R AS A UNIT.** To remove the primary pump as shown in Fig. IH644, first remove glass bowl from water trap and disconnect the fuel lines from the pump. Remove the two cap screws which re-

Fig. IH643—IH injection pump drive gear and timing pointer installation.

A. Chisel marked groove	C. Hub
B. Cap screws	D. Timing pointer

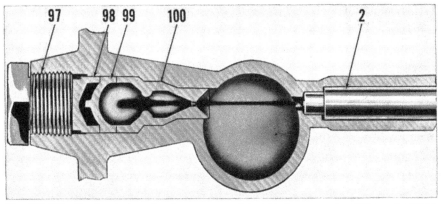

Fig. IH642J—Sectional view showing energy cell installation on series 350D and 350DU. Cell (100) is internally threaded for a puller screw.

2. Injector	97. Plug	98. Retainer	99. Cell cap

tain the primary pump to the injection pump and remove the pump.

Install the primary pump by reversing the removal procedure, making certain that the holes in gasket (2) are properly aligned with holes in the injection pump housing. Tighten the two pump retaining cap screws to a torque of 20-23 ft.-lbs. Bleed the fuel system as outlined in paragraph 99.

108. **PLUNGER—R&R AS A UNIT.** To remove the plunger unit (D—Fig. IH645), first clean dirt from the injection pump, fuel lines and connections and remove the pressure gage from the fuel inlet fitting. Remove the fuel return pipe and washers and remove the injection pump oil filler pipe. Remove the nut from each injection pump discharge fitting and remove the housing cover. Remove the

high pressure pipe (E), cap screws (F) and lift the plunger unit from the injection pump.

To install the plunger unit, turn the control gear so that the double-width tooth space is centered in the cutaway portion of the plunger bushing retainer and insert plunger unit so that the double width tooth of the rack will engage the double width tooth space of the gear. See Fig. IH646. Install the plunger retaining cap screws and check for binding by moving the rack through full range of travel. Slight binding condition can be corrected by shifting the plunger in its mounting. Install the high pressure line and tighten the retaining cap screws to a torque of 35-40 ft.-lbs. Install the remaining parts by reversing the removal procedure and tighten the discharge fitting clamp nuts to a

torque of 7-10 ft.-lbs. Tighten the return pipe cap to a torque of 80 ft.-lbs. Bleed the fuel systems as outlined in paragraph 99.

109. **DISTRIBUTOR BLOCK—R&R AS A UNIT.** To remove the distributor block unit (G—Fig. IH645), remove the pump pressure gage, fuel return pipe and the injection pump oil filler pipe. Remove the nut from each injection pump discharge fitting and remove the housing cover. Remove the high pressure pipe (E) and discharge fitting locks (H). Using a special tool, remove the four discharge fittings (J). Remove the distributor block retaining cap screws and lift the distributor block from the pump housing. When installing the discharge fittings, tighten them to a torque of 30 ft.-lbs.

If the old style distributor block gasket is used during reassembly, be sure that the elongated cut-out in gasket connects the drilled passages in the pump housing. New style gasket has cut-outs at both ends and cannot be installed incorrectly. Install the distributor block and tighten the retaining cap screws to a torque of 20 ft.-lbs. Tighten the high pressure line cap screws to a torque of 35-40 ft.-lbs. Install the housing cover, tighten the discharge fitting clamp nuts to a torque of 7-10 ft.-lbs. and the return pipe cap to a torque of 80 ft.-lbs. Bleed the fuel system as outlined in paragraph 99.

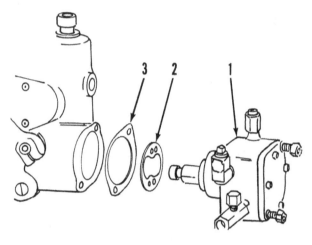

Fig. IH644—IH primary pump unit (1) removed from the injection pump.

2. Bearing cage gasket
3. Housing gasket

Fig. IH645—Side view of IH injection pump with the housing cover removed.

D. Plunger unit	G. Distributor block	X. Index notches
E. High pressure pipe	H. Locks	50. Lever
F. Cap screw	J. Discharge fitting	51. Shaft

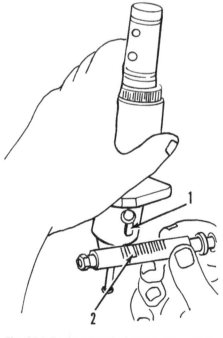

Fig. IH646—Install IH plunger unit so that double width tooth space (1) in gear meshes with double width tooth (2) on rack.

110. GOVERNOR-ADJUST. To adjust the governor, first remove the pump housing oil filler neck and loosen lock nut (1—Fig. IH647) on the rack. Refer to table X for governor adjustment specifications.

Remove the pump housing side cover and adjust the torque lever stop screw (10—Fig. IH648) until the face of the screw which contacts the torque lever stop pin (11) is dimension (A) from the finished surface of the torque lever (12). Adjust the torque shoe screw (13) so the lower face of the shoe is dimension (B) from finished surface beneath the lock nut. Adjust the overload gap screw (14) to give dimension (C) between the torque lever stop screw (10) and the torque lever stop pin (11). With the speed control lever in full shut-off position, adjust stop screw (2) until there is $\frac{1}{32}$ to $\frac{3}{32}$-inch clearance between the pump housing and the forward end of the torque control assembly.

Remove one injection nozzle and reconnect the nozzle to the nozzle tube. With the speed control rod in closed position, start engine and run on gasoline cycle. Turn the rack adjusting nut (3) until injection just starts. Then turn the nut back until injection stops. To insure positive shut-off, turn the nut 1½ additional turns in the same direction (direction which stopped injection). Adjust the engine high idle speed to 1580 rpm with stop screw (4).

With engine running at high idle speed, adjust the high idle gap between torque arm stop screw (15) and torque arm (16) to dimension (D) by turning stop screw (15). Adjust the bumper spring stop (17) to give a slight pressure on the bumper spring

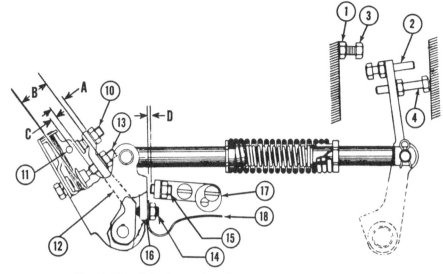

Fig. IH648—Side view of IH injection pump governor linkage.

10. Torque lever stop screw	13. Torque shoe screw	16. Torque arm
11. Torque lever stop pin	14. Overload gap screw	17. Bumper spring stop
12. Torque lever	15. High idle gap stop screw	18. Bumper spring

(18) at high idle speed. If the engine surges, increase the bumper spring pressure slightly. Recheck the torque lever pick up screw adjustment, high idle speed and make certain that engine will shut off on the diesel cycle.

Note: The tractor should now be ready to test under actual load conditions (as if working in the field).

If engine surges at high idle speed, increase the pressure on the bumper spring by adjusting the bumper spring stop.

If engine lacks power at rated load speed and overload speed, increase the gap between the torque arm (16) and stop screw (15) by backing out the screw. Do not adjust the stop screw to a point where exhaust smoke is excessive. Conversely, if smoke and power are both excessive, decrease the gap.

If engine tends to stall easily and lacks power at overload speeds, back out screw (14). Do not adjust the screw to a point where exhaust smoke is excessive.

If engine performance is satisfactory except for excessive smoke at overload speeds, turn the screw (14) in slightly

If tractor seems to lack power at rated load speed, but power is O. K. at overload speeds, adjust the high speed stop screw (4) to permit pulling the control lever farther back.

ROOSA-MASTER INJECTION PUMP

Series 350D-350DU

Series 350D tractors have a fully loaded engine speed of 1750 rpm and are fitted with Roosa-Master injection pump No. D/GV-C-L-435-27C. Series 350DU tractors have a fully loaded engine speed of 2000 rpm and are fitted with Roosa-Master injection pump No. D/GV-C-L-435-17C.

The subsequent paragraphs will outline ONLY the service work which can be accomplished without disassembling the pump. If pump requires complete overhaul, it should be turned over to an official Diesel service station. Inexperienced service personnel

Fig. IH647—Top view of IH injection pump with oil filler neck, removed.

1. Lock nut	3. Adjusting nut
2. Shut-off stop screw	4. High speed stop screw

TABLE X
DIESEL ENGINE GOVERNOR ADJUSTMENTS

	400D & W400D Prior to Engine Serial 21481	400D & W400D After Engine Serial 21480	450D and W450D
(A) Torque Lever Stop Screw	7/16-inch	7/16-inch	7/16-inch
(B) Torque Spring Shoe Screw	5/8-inch	5/8-inch	19/32-inch
(C) Overload Gap	0.045	0.045	0.025
(D) High Idle Gap	0.081	0.081	0.085
(E) Engine High Idle Speed, rpm	1580	1580	1580
(F) Engine Full Load Speed, rpm	1450	1450	1450
(G) Engine Low Idle Speed, rpm	600-630	600-650	600-650

should never attempt to overhaul a Diesel injection pump.

110A. **TIMING.** To time the injection pump, turn crankshaft clockwise (viewed from front) until number one piston is coming up on compression stroke and continue turning the crankshaft until pointer extending from timing gear cover is in register with the first notch (timing notch) on crankshaft pulley. The second notch indicates TDC. Refer to Fig. IH648A. Remove timing hole cover (Fig. IH-648B) from side of injection pump and check to be sure that the timing lines on drive plate, cam and pump housing are aligned as shown in Fig. IH-648C. If timing marks are not aligned, loosen the two pump mounting nuts, turn the pump housing either way as required to align the marks and tighten the mounting nuts.

110B. **INJECTION PUMP — R&R.** To remove the complete injection pump unit, first shut off the fuel supply and thoroughly clean dirt from pump, fuel lines and connections. Turn crankshaft clockwise (viewed from front) until number one piston is coming up on compression stroke and continue turning the crankshaft until pointer extending from timing gear cover is in register with the first notch (timing notch) on crankshaft pulley as shown in Fig. IH648A. Disconnect the fuel lines and controls, remove the two pump mounting nuts and withdraw pump from engine.

The injection pump is driven from the engine camshaft via the slotted coupling (Fig. IH648D). Before installing the injection pump, move the coupling back and forth and check the backlash between the coupling gear and the camshaft pinion. Recommended backlash is 0.004-0.006. If backlash is not as specified, lift the

Fig. IH648B—Series 350D and 350DU injection pump timing marks can be viewed after removing the timing hole cover. To change timing, loosen nuts and turn pump.

eccentric bushing up off of the locating pin, turn the bushing one serration, push the bushing downward and recheck the backlash. Turning the bushing clockwise decreases the backlash.

Before reinstalling the pump, remove the timing hole cover from side of injection pump and turn the pump drive shaft until timing lines are aligned as shown in Fig. IH648C. Mount pump on engine and connect fuel lines and controls.

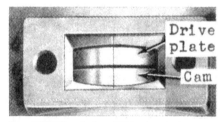

Fig. IH648C—Series 350D and 350DU injection pump timing marks as seen when pump timing hole cover is removed.

Recheck the pump timing as in paragraph 110A and bleed the fuel system as in paragraph 99C.

110C. **GOVERNOR - ADJUST.** To adjust the governor, first start engine and bring to normal operating temperature. Move the speed control hand lever to the wide open position, loosen the jam nut and turn the high speed adjusting screw (1 — Fig. IH648E) either way as required to obtain an engine high idle no-load speed of 1925 rpm for series 350D, 2200 rpm for series 350DU. Tighten the adjusting screw jam nut.

Note: With the high idle speed properly adjusted, the full load engine speed will be 1750 rpm for series 350D, 2000 rpm for series 350DU. Move the speed control hand lever to the low idle speed position, loosen the jam nut and turn adjusting screw (2) either way as required to obtain an engine slow idle speed of 600 rpm. Tighten the adjusting screw jam nut.

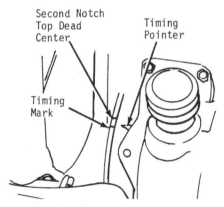

Fig. IH648A—When timing the injection pump on series 350D and 350DU, the pointer on timing gear cover must be in register with first notch on crankshaft pulley.

Fig. IH648D—The 350D and 350DU injection pump is driven from the engine camshaft by the coupling. Backlash between the coupling gear and camshaft gear should be 0.004-0.006.

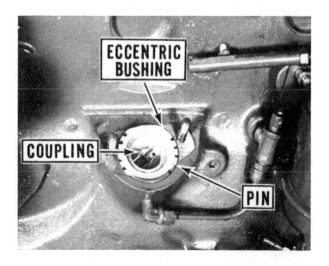

Fig. IH648E—Roosa-Master injection pump installation, showing the location of governor adjusting screws.

1. High speed screw
2. Low speed screw
3. Fuel stop screw
4. Run position screw

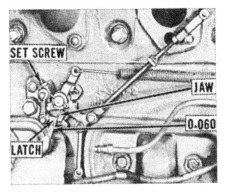

Fig. IH651—Checking the clearance between the starting linkage cross shaft latch and the cross shaft jaw.

Screws (3 and 4) are provided to set the limits of shut-off arm travel and should not normally require adjustment in the field. Screw (3) adjusts for maximum travel toward the "shut-off" position; whereas, screw (4) adjusts for "run" position. Adjustment of either screw requires removal of control cover and should be done only by experienced diesel service personnel.

DIESEL STARTING SYSTEM

Series 400D-W400D-450D-W450D

International Diesel engines are started on ordinary gasoline after first closing the Diesel engine throttle and pulling the compression release lever. Pulling the compression release lever accomplishes four things:

A starting valve in each engine cylinder is opened, thereby enlarging the combustion chamber and reducing the compression ratio to approximately that of an ordinary gasoline engine. It closes two butterfly valves in the Diesel air intake manifold and allows air to pass through the starting carburetor. The magneto or battery ignition electrical circuit is completed by opening of the grounding switch which is located in the forward portion of the intake manifold. It releases the float in carburetor, allowing the carburetor inlet needle to be actuated by the float.

After the engine runs on gasoline approximately one minute, the engine is switched over to full Diesel operation by releasing the compression release mechanism and at the same time opening the Diesel engine throttle. Releasing the compression release mechanism opens the manifold butterfly valves, closes the cylinder head starting valves, locks the carburetor needle valve on its seat and closes the magneto or battery ignition grounding switch.

111. **ADJUSTMENTS.** If the starting controls have been removed, or if erratic action occurs, inspect all parts for wear and renew any which are questionable. Remove the valve cover

and complete manifold and adjust the mechanism as follows:

Set the starting controls for diesel operation and proceed as follows: Adjust the length of the operating rod until the center to center distance between attaching pin holes is $6\frac{5}{16}$ inches as shown at (X) in Fig. IH649. Check the cross shaft end play which should be 0.030, measured between cross shaft bracket and cross shaft operating lever as shown in Fig. IH650 If the end play is not as specified, loosen the lever on each end of the cross shaft and shift the parts as required; then recheck the end play after tightening the lever clamp bolts.

Loosen the locknuts and turn the set screw in cross shaft latch bracket until there is 0.060 clearance between the cross shaft latch and the cross shaft jaw as shown in Fig. IH651. Turn the cross shaft clockwise until the pick-up faces on the cross shaft jaw and the cross shaft jaw levers are just contacting as shown in Fig. IH-652. Then, turn the screw in the cross shaft bracket until there is a clearance of 0.100 between screw and cross shaft operating lever. Tighten the

Fig. IH649—On series 400D, W400D, 450D and W450D, adjust the length (X) of the operating rod to 6 5/16 inches.

Fig. IH650—Checking the end play of series 400D, W400D, 450D and W450D starting linkage cross shaft.

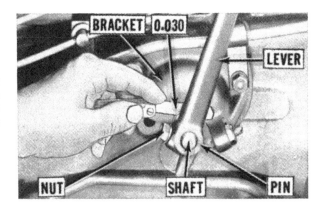

locknut and recheck the adjustment as shown in Fig. IH653.

Switch the controls to the gasoline starting position and using a screw driver as shown in Fig. IH654, push down on the starting valve covers to make certain that each cover has an additional downward travel of 1/64-inch before bottoming in the cylinder head. If the closest fitting cover does not have at least 1/64-inch additional downward travel, adjust the length of the operating rod (Fig. IH649) in increments of ½ turn until the proper travel is obtained.

Adjust the set screw in cross shaft jaw to obtain 0.015 clearance between adjusting screw and stop on cross shaft latch bracket as shown in Fig. IH655.

Note: With the preceding adjustments completed, and the controls in the gasoline starting position, again check to make certain that each of the starting valve covers have the aforementioned 1/64-inch additional travel before bottoming in cylinder

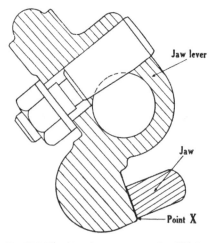

Fig. IH652—Showing contact point (X) between the cross shaft jaw and jaw lever. With parts in this position, make the adjustment shown in Fig. IH653.

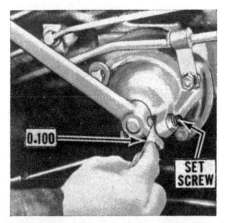

Fig. IH653—Checking clearance between the cross shaft release lever and the bracket set screw.

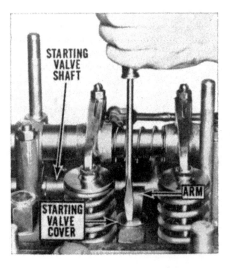

Fig. IH654—Checking the starting valve cover travel on series 400D, W400D, 450D and W450D engines. With the controls in the starting position, the covers should have an additional downward travel of 1/64-inch when manually pushed down as shown.

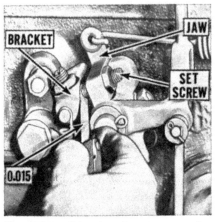

Fig. IH655—Checking the clearance between the set screw in the cross shaft jaw and the latch bracket.

head. If this recheck shows insufficient travel, all controls must be rechecked to locate possible adjustment errors.

Trip the controls to diesel position and check to be sure there is some clearance between the starting valve shaft arms and the starting valve covers. Refer to the following paragraph before installing the intake manifold.

112. INTAKE MANIFOLD. The battery ignition or magneto grounding switch and the two butterfly valves which close off the diesel air intake passages during the gasoline cycle are located in the intake manifold. The linkage which is shown exploded from the manifold in Fig. IH656 must be free with absolutely no tendency to bind. Adjust the butterfly valves (6) with set screws in top of manifold so that valves are perfectly horizontal inside the air passages when the levers on the shafts are against the set screws. The butterfly valves must fit the manifold with a maximum of 0.0015 clearance when measured with a ⅛-inch wide feeler gage. Be sure there is good contact of the ignition grounding switch. There should be no spark at the spark plugs when engine is running on the diesel cycle.

AIR RESTRICTION VALVE
Series 350D-350DU

113. If engine does not idle smoothly, remove plug from intake manifold and connect a vacuum gage. With engine running at 600 rpm, loosen the jam nut and turn the adjusting screw (Fig. IH657 or 658) either way as required to provide a manifold vac-

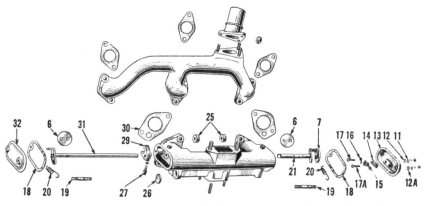

Fig. IH656—Exploded view of series 400D, W400D, 450D and W450D engine manifolds and associated parts.

6. Butterfly valves
7. Contact disc
11. Terminal screw washer
12. Terminal insulator
13. Manifold front end cover
14. Switch blade
    mounting strip
15. Switch contact blade
16. Contact blade
17. Terminal screw
18. Gasket
19. Stud
20. Manifold control spring
21. Manifold shaft
25. Control shaft seal
26. Expansion plug
27. Control lever pin
29. Control lever
30. Gasket
31. Control shaft
32. Rear cover

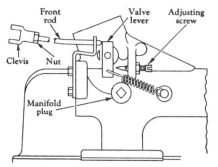

Fig. IH657—Manifold air restriction valve adjustments for series 350D.

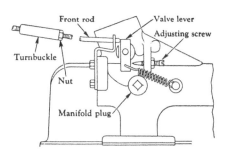

Fig. IH658—Manifold air restriction valve adjustments for series 350DU.

uum of 17 inches of mercury. In lieu of a vacuum gage, an approximate setting can be obtained by turning the adjusting screw either way as required to obtain a smooth idle without excessive smoking at 600 rpm. Tighten the jam nut. Adjust the restriction valve front rod at clevis (Fig. IH657) or at turnbuckle (Fig. IH658) to provide 0.010 clearance between front end of rod and air restriction valve lever. Then tighten the jam nut.

NON-DIESEL GOVERNOR

International Harvester Governors are of the centrifugal flyweight type and are driven by a gear in the timing gear train. Before attempting any governor adjustments, check the operating linkage for binding condition or lost motion and correct any undesirable conditions.

Series 300-300U-350-350U-400-W400-450-W450

123. ADJUST. Adjust the carburetor; then with engine stopped, remove cover from side of governor housing. Turn the governor screw (10—Fig. IH664) until it just touches its stop; at which time, there should be no tension on governor spring (2). Place operator's hand lever in full speed position and adjust high idle speed screw (11) until it just touches stop (12). If length of threads on screw (11) will not permit this adjustment, lengthen or shorten the linkage between operator's hand lever and governor lever (3) until the screw is brought into range.

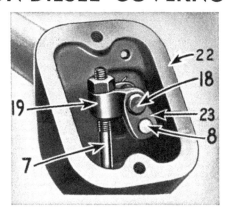

Fig. IH665—Synchronizing governor to carburetor linkage on non-Diesel models.

7. Connecting rod
8. Throttle shaft
18. Connecting rod pin
19. Adjusting block
22. Connecting rod housing
23. Throttle shaft lever

To adjust governor to speeds which follow, start engine and place operator's hand lever in full speed position; turn high idle speed screw (11) in to decrease speed or out to increase speed.

If correct speed adjustment cannot be obtained by the above procedure or if governor is just being installed on engine, adjust the governor to carburetor linkage as follows: With engine stopped, place operator's hand lever in full speed position and remove cover from top slanted surface of connecting rod housing (22—Fig. IH665). Remove the pin (18) and pull both the throttle shaft lever and the adjusting block up as far as they will go; then, with the parts thus held, it should be possible to freely insert pin

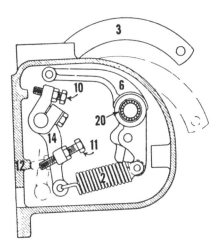

Fig. IH664—Non-Diesel governor housing with side cover removed, showing the adjustment points.

2. Governor spring
3. Speed change lever
6. Rockshaft lever
10. Idle stop screw
11. Speed adjusting screw
12. Screw stop
14. Governor spring lever
20. Rockshaft lever bearing

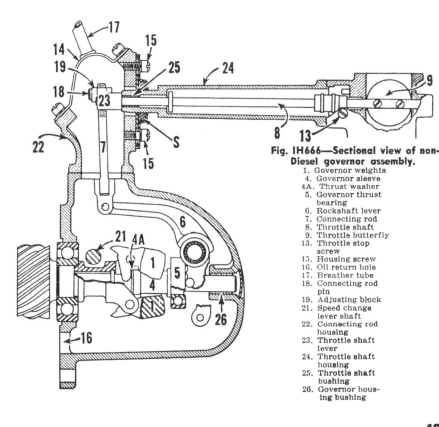

Fig. IH666—Sectional view of non-Diesel governor assembly.
1. Governor weights
4. Governor sleeve
4A. Thrust washer
5. Governor thrust bearing
6. Rockshaft lever
7. Connecting rod
8. Throttle shaft
9. Throttle butterfly
13. Throttle stop screw
15. Housing screw
16. Oil return hole
17. Breather tube
18. Connecting rod pin
19. Adjusting block
21. Speed change lever shaft
22. Connecting rod housing
23. Throttle shaft lever
24. Throttle shaft housing
25. Throttle shaft bushing
26. Governor housing bushing

(18). If pin does not enter freely, screw the block (19) on or off connecting rod (7) until this condition is obtained. The carburetor throttle butterfly is now synchronized to the governor, with governor in wide open position.

Recommended governed speeds are as follows:

Low Idle .... 425 rpm
High Idle: (No Load)
  Series 300-350 .... 1925 rpm
  Series 300U-350U .... 2200 rpm
  Series 400-W400-
    450-W450 .... 1600 rpm
Engine Fully Loaded rpm:
  Series 300-350 .... 1750 rpm
  Series 300U-350U .... 2000 rpm

Series 400-W400-
  450-W450 .... 1450 rpm

124. **R&R AND OVERHAUL.** To remove the unit from engine, loosen carburetor mounting screws and remove cap screws which hold front face of governor housing to the engine. Also remove ventilating tube (17—Fig. IH-666) and screws (15) from throttle shaft housing retainer. Pull throttle shaft housing (24) forward far enough to release throttle shaft (8) from the coupling at carburetor butterfly valve. When reinstalling governor unit, be sure throttle shaft fully engages coupling at carburetor. Allow throttle shaft and housing to find correct alignment before tightening screws (15).

The governor shaft forward ball bearing is retained in housing by a snap ring (not shown) which also takes the thrust of the helical drive gear. Check clearance of governor weights (1) on their hinge pins and freedom of governor sleeve (4) on governor shaft. The bushing (26) at rear end of shaft is renewable after removing the Welch plug from body. The ball thrust bearing (5) should be installed on sleeve (4) with the closed face away from the shoulder on sleeves. Needle bearings supporting the governor lever (6) are caged units. Linkage between governor lever (6) and carburetor should operate without binding.

COOLING SYSTEM

RADIATOR

Series 300-350-350D-400-400D-450-450D

125. To remove the radiator, drain cooling system and remove the hood and grille. Disconnect the radiator hoses and radiator upper support rod or bracket; then remove the dust shield from under front frame rails. Disconnect the shutter control rod. Remove the steering worm shaft, unbolt radiator from bolster and lift radiator from tractor.

Series 300U-350U-350DU-W400-W400D-W450-W450D

126. To remove the radiator, drain cooling system and remove the hood. Disconnect head light wires and re-

move grille. Disconnect the radiator upper support rod or bracket. On models so equipped, disconnect the shutter control rod and radiator hoses. Unbolt and remove radiator from tractor.

FAN

All Models

128. The fan is mounted on and driven by the water pump. To remove the fan blades on all models except the 350D and 350DU, it is first necessary to remove the radiator as outlined in paragraphs 125 or 126. On series 350D and 350DU, the fan blades can be removed without removing the radiator.

WATER PUMP

Series 300-300U-350-350U-400-400D-W400-W400D-450-450D-W450-W450D

129. **REMOVE AND REINSTALL.** To remove water pump from tractor, it is first necessary to remove radiator as outlined in paragraphs 125 or 126.

130. **REPACK.** Packing is furnished in split segments ¼-inch thick. To renew the packing (19—Fig. IH667 or Fig. IH668), it is advisable to first remove the pump unit from the engine as per paragraph 129. Remove driver pin (5) and driver (6). Unscrew packing nut (3), remove old packing and replace with new. Impeller and

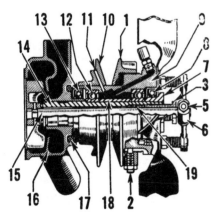

Fig. IH667—Sectional view of the water pump assembly which is typical of that used on series 300, 300U, 350 and 350U.

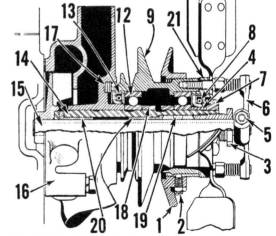

Fig. IH668—Sectional view of the water pump assembly which is typical of that used on series 400, 400D, W400, W400D, 450, 450D, W450 and W450D.

1. Pulley flange	5. Driver pin	10. Fixed flange
2. Flange set screw	6. Pump shaft driver	11. Flange retaining pin
3. Packing nut	7. Bearing retaining nut	12. Fan hub bearing
4. Bearing lock sleeve	8. Oil seal	13. Rear oil seal
	9. Fan hub	14. Pump shaft sleeve

15. Shaft with impeller and thrust washer	18. Bearing spacer
16. Pump body	19. Pump packing
17. Felt washer	20. Pump shaft bushings
	21. Bearing retainer

Fig. IH668A — Exploded view of water pump used on series 350D and 350-DU. Pump can be overhauled without removing body from engine.

6. Gasket
7. Impeller
8. Seal assembly
9. Drive support
10. Pump shaft and bearing assembly
11. Snap ring
12. Fan pulley
13. Retainer ring

shaft (15) may be removed for easy renewal of packing.

131. **OVERHAUL.** Disassemble the unit in the following manner: Remove driver pin (5—Fig. IH667 or Fig. IH-668), driver (6), packing nut (3), bearing retaining nut (7) and the front oil seal (8). The impeller and shaft (15) can now be withdrawn. Support the unit on rear flange of pulley hub (9—Fig. IH668) or fixed flange (10—Fig. IH667) and press on forward end of sleeve (14) which will release rear bearing (12) and pump body (16) from the fan hub.

On the pump shown in Fig. IH668, the adjustable pulley flange (1) can be removed by unscrewing after first removing the set screw (2). On the pump shown in Fig. IH667, the adjustable pulley flange (1) is removed in the same manner although it is first necessary to remove the fixed pulley flange pin (11) and unscrew the fixed pulley flange (10) from the hub.

On the series 300, 300U, 350 & 350U, if bushings in bore of sleeve (14—Fig. IH667) are worn it will be necessary to renew the sleeve assembly, as bushings are not furnished separately. On other models, the sleeve bushings (20—Fig. IH668) are furnished separately if desired. The impeller, thrust washer and shaft (15) are furnished only as a single unit. Impeller shaft diameter is 0.4355-0.436 for series 300, 300U, 350 and 350U, 0.6215-0.622 for other models. Clearance of shaft in bushings should be 0.0015-0.0025.

To reassemble pump, press sleeve (14) into pump body (16), making sure that the ⅛-inch hole in sleeve registers with a similar hole in body. Assemble adjustable flange (1) and fixed flange (10—Fig. IH667 only) to hub. Install rear bearing (12) and oil seal (13) in fan hub (9) with lip of seal facing bearing. Insert felt (17) and press the hub assembly on the sleeve and body unit while supporting on end of sleeve. Install spacer (18), front bearing, oil seal (8) with lip fac-

ing fan, lock sleeve (4—Fig. IH668 only) and bearing retaining nut (7).

Ball bearings are not preloaded by the spacer (18) so that when fully assembled the front bearing need not be in contact with shoulder in hub. If fan pulley does not revolve freely after bearing retaining nut (7) is fully tightened, the probable cause is that spacer (18) is too short, causing the bearings to be preloaded. Correct this with a new spacer.

Reinstall impeller, thrust washer and shaft (15), packing (19) and packing nut (3). Install bearing retainer (21—Fig. IH668 only), fan blades, driver (6) and driver pin (5).

Series 350D-350DU

131A. Removal of water pump on series 350DU requires R&R of radiator. On series 350D, the pump can be removed without R&R of radiator. To overhaul the pump, proceed as follows: Refer to Fig. IH668A.

Unbolt drive support (9) from pump body and remove the drive support assembly. Using a suitable puller, remove impeller (7) and seal assembly (8). Remove retainer ring (13) and pulley (12). Extract snap ring (11) and press the shaft and bearing assembly out of the drive support.

Inspect all parts and renew any which are damaged or worn. Assemble pump by reversing the disassembly procedure. Install the drive housing assembly so that drain hole in same is down.

WATER HEADER PLATES
Series 300-300U-350-350U-400-400D-W400-W400D-450-450D-W450-W450D

132. Non-Diesel engines are equipped with one water header plate and Diesel engines are equipped with two plates. To remove the front plate, on Diesels, it is necessary to R&R injection pump as outlined in paragraph 106. To remove the rear plate, it is necessary to R&R the fuel filters.

IGNITION AND ELECTRICAL SYSTEM

DISTRIBUTOR TIMING
Series 300-300U-350-350U-400-400D-W400-W400D-450-450D-W450-W450D

133. To install and time the battery ignition unit, first crank engine until No. 1 (front) piston is coming up on compression stroke and continue cranking slowly until the (DC) notch on the fan drive pulley is in register

with the pointer extending from front face of timing gear cover as shown in Fig. IH669 for series 300 and 350, Fig. IH670 for series 300U and 350U, Fig. IH671 for series 400, W400, 450 and W450, and Fig. IH672 for series 400D, W400D, 450D and W450D.

Turn the distributor drive shaft until rotor arm is in the No. 1 firing position and mount the ignition unit on

Top Dead Center

16° Advance

22° Advance

30° Advance

Fan Drive Pulley

Timing Pointer

Fig. IH669—Fan pulley timing marks for series 300 and 350. Refer to paragraph 135A or 135B for explanation of timing notches.

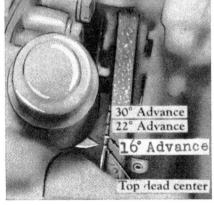

30° Advance
22° Advance
16° Advance
Top dead center

Fig. IH670—Fan pulley timing marks for series 300U and 350U. Refer to paragraph 135A for explanation of timing notches.

the engine, making certain that lugs (A—Fig. IH673) on ignition unit engage slots (B) in the drive coupling.

Note: If the driving lugs will not engage the coupling drive slots, when rotor is in No. 1 firing position, it will be necessary to remesh the drive gears as follows: Grasp the distributor drive shaft and pull same outward to disengage the gears. Turn drive shaft until lugs will engage the drive slots, then push drive shaft inward to engage the gears.

134. To time the battery ignition distributor after same is installed as previously outlined, proceed as follows: Loosen the distributor mounting cap screws and retard distributor by turning distributor assembly in same direction as the cam rotates. (Refer to Table 2 for distributor rotation). Disconnect coil secondary cable from distributor cap and hold free end of cable $\frac{1}{16}$-$\frac{1}{8}$-inch from distributor primary terminal. Advance the distributor by turning the distributor body in opposite direction from cam rotation until a spark occurs at the gap. Tighten the distributor mounting cap screws at this point. Assemble the spark plug cables to the distributor cap in the proper firing order of 1-3-4-2.

Running timing on non-Diesel models can be checked with a neon-type timing light using the advance data given in Table 2 and the advance notches located on flange of fan drive pulley (Figs. IH669, IH670 and IH671).

FAN DRIVE PULLEY TIMING MARKS

135. Fan drive pulleys are equipped with notches indicating TDC as well as the fully advanced timing point for the various distributors used. Series 300, 300U, 350 and 350U pulleys may be equipped with three or four notches and series 400, W400, 450 and W450

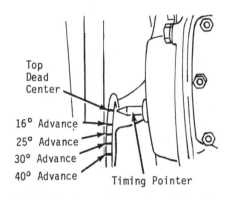

Fig. IH671—Fan pulley timing marks for series 400, W400, 450 and W450. Refer to paragraph 135C or 135D for explanation of timing notches.

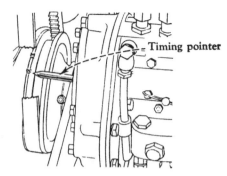

Fig. IH672—Fan pulley timing marks for series 400D, W400D, 450D and W450D.

may be equipped with either four or five notches. An explanation of the notches is as follows:

Series 300-300U-350-350U With 3 Notches

135A. The first notch indicates 30 degree advance and is used with symbol "J" distributors on distillate or kerosene burning engines. The second notch indicates 22 degrees advance and is used with symbol "O" or "S" distributors on gasoline burning engines. The third notch indicates top dead center for all distributors.

Series 300-300U-350-350U With 4 Notches

135B. The first notch indicates 30 degrees advance and is used with symbol "J" distributors on distillate or kerosene burning engines. The second notch indicates 22 degrees advance and is used with symbol "O" or "S" distributors on gasoline burning engines. The third notch indicates 16 degree advance and is used with symbol "T" distributors on LP-gas burning engines. The fourth notch indicates top dead center for all distributors.

Series 400-W400-450-W450 With 4 Notches

135C. The first notch indicates 40 degree advance and is used only when this pulley is used as a service pulley on C-248 engines with a symbol "A" distributor. This first notch is NOT used in timing any of the 400-W400-450-W450 series tractors. The second notch indicates 30 degree advance and is used with symbol "J" distributors on distillate or kerosene burning engines. The third notch indicates 25 degree advance and is used with symbol "N" distributors on gasoline burning engines. The fourth notch indicates top dead center for all distributors.

Series 400-W400-450-W450 With 5 Notches

135D. The first notch indicates 40 degree advance and is used only when this pulley is used as a service pulley on C-248 engines with a symbol "A" distributor. This first notch is NOT used in timing any of the 400, W400, 450 or W450 series tractors. The second notch indicates 30 degree advance and is used with symbol "J" distributors on distillate or kerosene burning engines. The third notch indicates 25 degree advance and is used with symbol "N" distributors on gasoline burning engines. The fourth notch indicates 16 degree advance and is used with symbol "T" distributors on LP-gas burning engines. The fifth notch indicates top dead center for all distributors.

DISTRIBUTOR OVERHAUL
All Models So Equipped

136. Defects in the battery ignition system may be approximately located by simple tests which can be performed in the field; however, complete ignition system analysis and component unit tests require the use of special testing equipment. Most of the distributors have automatic spark advance which is obtained by a centrifugal governor built into the unit.

Distributor identification symbol is the first letter of the code stamped on the outside of the distributor mounting flange. Point setting should be 0.020 and breaker arm spring pressure should be 21-25 ounces for all distributors. Refer to the following specification data which is listed by the various distributor symbols.

Symbol "C"

Rotation viewed from
   driving end ....CCW
Advance data........No advance, fixed spark design

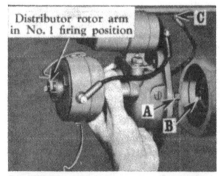

Fig. IH673—When installing the battery ignition unit, make certain that lugs (A) engage slots (B) in the drive coupling.

Symbol "H"
Rotation viewed from
driving end ....CCW
Advance data (distributor degrees
@ distributor rpm)
Start advance .... 0 @ 200
Intermediate ....2.5 @ 400
Full advance ....4 @ 462.5

Symbol "J"
Rotation viewed from
driving end ....CW
Advance data (distributor degrees
@ distributor rpm)
Start advance .... 0 @ 200
Intermediate ....4.5 @ 400
Intermediate ....12 @ 600
Full advance ....15 @ 800

Symbol "N"
Rotation viewed from
driving end ....CW
Advance data (distributor degrees
@ distributor rpm)
Start advance .... 0 @ 200
Intermediate ....4.5 @ 400
Intermediate ....9.5 @ 600
Full advance ....12.5 @ 800

Symbol "O"
Rotation viewed from
driving end ....CW
Advance data (distributor degrees
@ distributor rpm)
Start advance .... 0 @ 200
Intermediate .... 4 @ 400
Intermediate .... 7 @ 600
Intermediate ....10 @ 800
Full advance ....11 @ 900

NOTE: Early 300U tractors were fitted with a symbol "O" distributor. When servicing these models, advance spring kit No. 364654 R91 should be installed to change advance curve to coincide with symbol "S" unit.

Symbol "S"
Rotation viewed from
driving end ....CW
Advance data (distributor degrees
@ distributor rpm)
Start advance .... 0-1 @ 200
Intermediate .... 3-5 @ 400
Intermediate .... 6-8 @ 600
Intermediate ......8.5-10.5 @ 800
Full advance .....11-11.5 @ 900

Symbol "T"
Rotation viewed from
driving end ....CW
Advance data (distributor degrees
@ distributor rpm)
Start advance .... 0 @ 200
Intermediate .... 4 @ 400
Intermediate .... 7 @ 600
Full advance .... 8 @ 700

Symbol "X"
Rotation viewed from
driving end ....CW

Advance data (distributor degrees
@ distributor rpm)
Start advance .... 0-1 @ 200
Intermediate ......4.5-6.5 @ 400
Intermediate ...... 9-11 @ 600
Intermediate ......13-15 @ 800
Full advance .....15-15.5 @ 900

GENERATOR, REGULATOR & STARTING MOTOR
All Models
137. Tractors are equipped with Delco-Remy electrical units. Refer to the actual unit for model number. Third brush spring tension should be 19 oz. on all generators so equipped. Specification data for generator, voltage regulator and starting motor units used are as follows:

GENERATORS
1100042, 1100052
Brush spring tension.........28 oz.
Field draw, volts.............. 6
amperes 1.85-2.03
Cold output, volts.............8.0
amperes 35
rpm 2650

1100305
Brush spring tension.........28 oz.
Field draw, volts..............12
amperes 1.58-1.67
Cold output, volts.............14.0
amperes 20
rpm 2300

1100501
Brush spring tension.........24 oz.
Field draw, volts.............. 6
amperes 2.5-2.72
Cold output, volts.............7.0
amperes 20-25
rpm 2400

1100531
Brush spring tension.........16 oz.
Field draw, volts.............. 6
amperes 2.5-2.72
Cold output, volts.............7.0
amperes 20-25
rpm 2400

1100964
Brush spring tension.........24 oz.
Field draw, volts..............12
amperes 2.0-2.14
Cold output, volts.............14.0
amperes 11-13
rpm 2300

1100996, 1100997, 1101001
Brush spring tension.........24 oz.
Field draw, volts..............12
amperes 2.0-2.14
Cold output, volts.............14.0
amperes 13-16
rpm 2600

VOLTAGE REGULATORS
1118779
Ground polarity ....Positive
Cut-out relay, air gap........0.020
point gap 0.020
closing voltage, range...11.8-14.0
adjust to 12.8
Voltage regulator, air gap.....0.075
setting volts, range......13.6-14.5
adjust to 14.0

1118780
Ground polarity ....Positive
Cut-out relay, air gap........0.020
point gap 0.020
closing voltage, range....5.9-7.0
adjust to 6.4
Voltage regulator, air gap.....0.075
setting volts, range.......6.6-7.2
adjust to 6.9

STARTING MOTORS
1107169
Volts 6
Brush spring tension (min.), oz...24
No load test, volts.............5.6
maximum amperes 80
minimum rpm 5500
Lock test, volts................3.0
maximum amperes 550
minimum torque, ft.-lbs......11

1108038, 1108041
Volts 6
Brush spring tension (min.), oz...24
No load test, volts.............5.67
maximum amperes 80
approximate rpm 5500
Lock test, volts................3.25
maximum amperes 600
minimum torque, ft.-lbs......14

1108140
Volts 12
Brush spring tension (min.), oz...24
No load test, volts.............11.3
maximum amperes 70
approximate rpm 6000
Lock test, volts................6.7
maximum amperes 530
minimum torque, ft.-lbs......16

1108626, 1108642, 1108646
Volts 12
Brush spring tension (min.), oz...24
No load test, volts.............11.8
minimum amperes 40
maximum amperes 70
minimum rpm 6800
maximum rpm 9200
Lock test, volts................5.85
maximum amperes 615
minimum torque, ft.-lbs......29

1108996
Volts 12
Brush spring tension, oz.....36-40
No load test, volts.............11.3
maximum amperes 65
approximate rpm 5500
Lock test, volts................4.0
maximum amperes 675
minimum torque, ft.-lbs......30

ENGINE CLUTCH

139. All models are fitted with single plate, spring loaded, dry disc type clutches manufactured by either International Harvester or Rockford. The two clutch cover assemblies are interchangeable and the adjustment and overhaul procedures are the same.

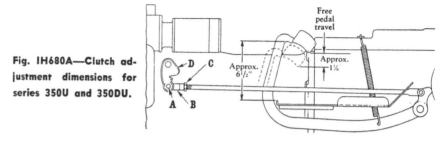

Fig. IH680A—Clutch adjustment dimensions for series 350U and 350DU.

ADJUSTMENT

Series 300U
(Without Torque Amplifier)

140. Adjustment to compensate for lining wear is accomplished by adjusting the clutch pedal linkage, not by adjusting the position of the clutch release levers.

To adjust the linkage, loosen bolt (A—Fig. IH680) and raise the clutch pedal as far as it will go; at which time, the distance from top of pedal pad to top of platform should be 7⅜ inches. If the measured distance is not 7⅜ inches, turn stop screw (B) either way as required until the specified measurement is obtained. Then, move adjusting plate (E) either way as required to obtain a pedal free travel of 1¼ inches and tighten bolt (A). Depress pedal to the extreme down position and measure the distance the pedal travels. Turn stop screw (C) either way as required until the measured pedal full travel is 5¼ inches.

Series 350U-350DU
(Without Torque Amplifier)

140A. Adjustment to compensate for lining wear is accomplished by adjusting the clutch pedal linkage, not by adjusting the position of the clutch release levers.

To adjust the linkage, loosen nut (C—Fig. IH680A), remove clevis pin (A) and turn clevis (B) either way as required to obtain a pedal free travel of 1⅛ inches as shown.

All Other Models
(Without Torque Amplifier)

141. Adjustment to compensate for lining wear is accomplished by adjusting the clutch pedal linkage, not by adjusting the position of the clutch release levers.

To adjust the linkage, loosen lock nut (C—Fig. IH681), remove clevis pin (A) and turn clevis (B) either way as required to obtain the specified free travel measured between pedal and pedal contact point on the rear frame cover. The specified pedal free travel is as follows:

Series 300 ....1¼ inches
Series 350-350D ....1¾ inches
Series 400-400D ....⅞-inch
Series W400-W400D 1⅟₁₆-inch
Series 450-450D ....1¼ inches
Series W450-W450D ....1⅓⁄₁₆-inch

On models so equipped turn the stop screw (E) either way as required to obtain the specified pedal full travel measured between pedal and pedal contact point on the rear frame cover. The specified pedal full travel is as follows:

Series 300 ....3¾ inches
Series 400-400D ....4 inches
Series W400-W400D ....2⅓⁄₁₆ inches

All Models
(With Torque Amplifier)

142. Adjustment to compensate for lining wear is accomplished by adjusting the clutch pedal linkage, not by adjusting the position of the clutch release levers.

The engine clutch linkage and the torque amplifier clutch linkage should be adjusted at the same time. The adjustment procedure is given in paragraphs 156, 156A or 157.

REMOVE AND REINSTALL
All Models

143. To remove the engine clutch, it is first necessary to detach (split) engine from clutch housing as outlined in paragraphs 144, 145 or 146. The clutch can then be unbolted and removed from flywheel in the conventional manner.

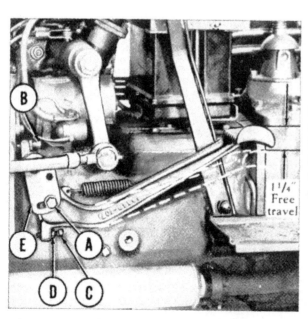

Fig. IH680—Side view of the 300 Utility clutch housing showing the location of adjustments for the engine clutch on models without torque amplifier.

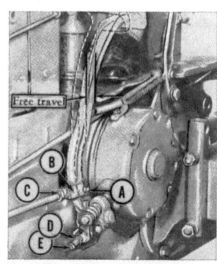

Fig. IH681—Clutch adjustment points for series 300, 400 and 400D. Except for details of construction, series W400, W400D, 350, 350D, 450, 450D, W450 and W450D are similar.

TABLE 3
CLUTCH OVERHAUL DATA

Tractor Models	Clutch Made By	(A) Cover Setting	(B) Lever Height	Pressure Springs Free Length	Color	Part No.	Lbs. Test	at Height
300-300U-350-350U-350D-350DU	I-H	0.851"	2 7/32"	2.828	White Stripe	IH 360 850 R1	104-118 @	1.44"
	I-H	0.851"	2 7/32"	2.75	....	IH 369 229 R1	133-147 @	1.44"
300-300U-350-350U-350D-350DU	Rockford	0.851"	2 7/32"	....	None	IH 364 327 R1	110-120 @	1.44"
	Rockford	0.851"	2 7/32"	2.75	....	IH 369 349 R1	133-143 @	1.44"
400-400D-W400-W400D-450-450D-W450-W450D	I-H	1.020"	2 5/16"	....	Orange	IH 364 329 R1	105-115 @	1 13/16"
	I-H	1.020"	2 5/16"	3.156	....	IH 369 231 R1	130-144 @	1.813"
400-400D-W400-W400D-450-450D-W450-W450D	Rockford	1.020"	2 5/16"	....	Orange	IH 364 329 R1	105-115 @	1 13/16"
	Rockford	1.020"	2 5/16"	3.156	....	IH 369 351 R1	130-140 @	1.813"

TRACTOR SPLIT
Series 300-350-350D-400-400D-450-450D

144. To detach (split) engine from clutch housing, first remove hood, then disconnect the steering shaft universal joint and unbolt the steering shaft center bearing from the fuel tank support. Disconnect the shutter control rod, drain cooling system and on models so equipped, disconnect the radiator upper support rod at rear. Disconnect the heat indicator sending unit, fuel lines, oil pressure gage line, wiring harness and controls from engine and engine accessories. Remove the filter and tool box. Disconnect hydraulic manifold from the hydraulic pump and on models with power steering, disconnect the hoses from the power steering unit. Attach hoist to engine half of tractor in a suitable manner and securely block rear half of tractor so it will not tip. Unbolt engine and side rails from clutch housing and separate the tractor halves.

Series W400-W400D-W450-W450D

145. To detach (split) engine from clutch housing, first remove hood, drain cooling system and disconnect the shutter control rod and head light wires. Disconnect the radiator upper support rod at rear and the steering drag link from steering (knuckle) arm. Disconnect the heat indicator sending unit, fuel lines, oil pressure gage line, wiring harness and controls from engine and engine accessories. Remove the oil filter and on Diesel models, remove the starting control rod, fuel supply and return lines and carburetor air intake pipe. Disconnect the hydraulic lines from the Hydra-Touch pump. On models equipped with power steering, unbolt the power steering pump from engine and without disconnecting hoses from pump, lay the pump rearward and out of way. Attach hoist to engine half of tractor in a suitable manner and securely block rear half of tractor so it will not tip. Unbolt engine and side rails from clutch housing and separate the tractor halves.

Series 300U-350U-350DU

146. To detach (split) engine from clutch housing, first drain cooling system and remove hood. Disconnect head light wires, radius rod pivot bracket from clutch housing and both drag links from the steering (knuckle) arms. Disconnect cable and remove starting motor. Disconnect the heat indicator sending unit, fuel lines, oil pressure line, wiring harness and controls from engine and engine accessories.

Attach hoist to engine half of tractor in a suitable manner and securely block rear half of tractor so it will not tip. Unbolt engine from clutch housing and separate the tractor halves. Note: The two long bolts retaining clutch housing to top of engine are unscrewed gradually as engine is moved forward. This procedure eliminates the need of removing the steering gear unit and usually saves considerable time.

OVERHAUL CLUTCH
All Models

147. A general procedure for disassembling, adjusting and/or overhauling the removed clutch cover assembly is given in the Rockford clutch section of the separate Standard Units Manual. Overhaul specifications are listed in Table 3. Dimensions (A) and (B) in the table are shown in Fig. IH682.

CLUTCH RELEASE BEARING
All Models

148. The procedure for renewing the clutch release bearing is evident after detaching engine from clutch housing as outlined in paragraphs 144, 145 or 146.

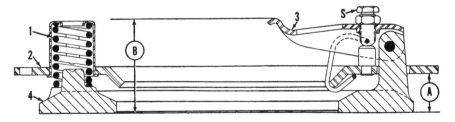

Fig. IH682—Sectional view of a Rockford spring loaded clutch cover, showing the release lever adjustment dimensions. (A) dimension is the position of the pressure plate in relation to the cover plate which must be maintained when adjusting the release lever height (B).

S. Lever adjusting screw   1. Pressure spring   2. Cover plate   3. Release lever   4. Pressure plate

CLUTCH SHAFT
All Models
(Without Torque Amplifier)

149. To remove the engine clutch shaft on models without torque amplifier, detach (split) engine from clutch housing as outlined in paragraphs 144, 145 or 146; then withdraw the clutch shaft forward and out of clutch hous-

Fig. IH683—Removing bearing cage (26), independent power take-off drive shaft (25) and clutch shaft (24) from front face of clutch housing.

TORQUE AMPLIFIER UNIT

Torque amplification is provided by a planetary gear reduction unit located between the engine clutch and the transmission. The unit is controlled by a hand operated, single plate, spring loaded clutch. When the clutch is engaged as in Fig. IH685, engine power is delivered to both the pri-

ing. The clutch shaft is shown installed in Figs. IH711 or IH716.

All Models
(With Torque Amplifier)

150. To remove the clutch shaft on all models with a torque amplifier, first detach (split) engine from clutch housing as outlined in paragraphs 144, 145 or 146 and remove the engine clutch release bearing and shaft. On models with independent power take-off, unbolt bearing cage (26—Fig. IH-683) from clutch housing and withdraw the independent power take-off drive shaft (25), bearing cage (26) and clutch shaft (24). Note: A tapped hole is provided in the end of the clutch shaft to aid in its removal.

On models with transmission driven power take-off, the engine clutch shaft and the power take-off drive gear are an integral unit (10A—Fig. IH690). To remove the shaft and gear, unbolt the bearing cage (5) from clutch housing and withdraw the bearing cage and shaft assembly. Remove shaft from bearing cage. The two shaft oil seals (one located in bearing cage and the other in the clutch housing) can be renewed at this time. The two ball type shaft carrier bearings can be pulled from shaft if renewal is required.

mary sun gear (PSG) and the planet carrier (PC). This causes the primary sun gear and the planet carrier to rotate as a unit and the system is in direct drive. When the clutch is disengaged as shown in Fig. IH-686, engine power is transmitted through the primary sun gear to the larger portion of

the compound planet gears (PG), giving the first gear reduction. The second gear reduction is provided by the smaller portion of the compound planet gears driving the secondary sun gear (SSG). As a result of the two gear reductions, an overall gear reduction of approximately 1.48:1 is obtained.

T. A. CLUTCH
All Models So Equipped

155. **ADJUST.** Adjustment to compensate for lining wear is accomplished by adjusting the clutch actuating linkage, not by adjusting the position of the clutch release levers. The engine clutch linkage and the torque amplifier clutch linkage should be adjusted at the same time. For series 300U, refer to the following paragraph 156; for series 350U and 350DU, refer to paragraph 156A; for other models, refer to paragraph 157.

156. SERIES 300U. Refer to Fig. IH-687, remove spring (E), loosen lock nut (F) and remove clevis pin (G). Loosen bolt (A) and raise the clutch pedal as far as it will go; at which time, the distance from top of pedal pad to top of platform should be 7⅜ inches. If measured distance is not 7⅜ inches, turn stop screw (B) either way as required until the specified measurement is obtained. Then, move adjusting plate (M) either way as required to obtain a pedal free travel of 1¼ inches as shown and tighten bolt (A). Depress pedal to the extreme down position and measure the distance the pedal travels. Turn stop screw (C) either way as required until the measured pedal full travel is 5¼ inches as shown.

After the engine clutch linkage is properly adjusted, place the torque amplifier control lever in the forward position as shown and proceed to ad-

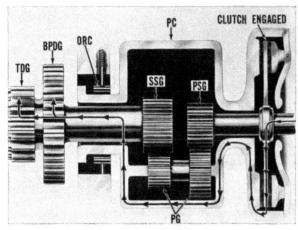

Fig. IH685—When the torque amplifier clutch is engaged, the system is in direct drive.

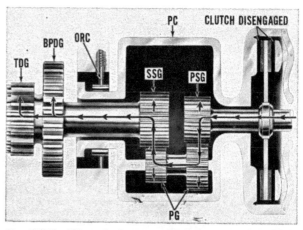

Fig. IH686—When the torque amplifier clutch is disengaged, an overall gear reduction of approximately 1.48:1 is obtained.

just the torque amplifier clutch linkage as follows:

Loosen lock nut (H), remove clevis pin (I) and turn lever (J) counterclockwise as far as possible without forcing. See inset. This places the T. A. clutch release bearing against the clutch release levers. Now, adjust clevis (K) to provide a space of $\frac{3}{16}$-inch between the inserted pin (I) and the forward end of the elongated hole in clevis (K). Tighten lock nut (H) and reinstall spring (E). Adjust the length of rod (N) with clevis (L) so that rod (N) is the shortest possible length that will not change the position of levers (M & J) when pin (G) is inserted.

156A. SERIES 350U-350DU. Refer to Fig. IH687A, remove spring (E), loosen lock nut (F) and remove clevis pin (G). Loosen nut (C), remove clevis pin (A) and turn clevis (B) either way as required to obtain a pedal free travel of $1\frac{1}{8}$ inches as shown.

After the engine clutch linkage is properly adjusted, place the torque amplifier control lever in the forward position as shown and proceed to adjust the torque amplifier clutch linkage as follows: Loosen lock nut (H), remove pin (I) and turn lever (J) counter-clockwise as far as possible without forcing. See inset. This places the T. A. clutch release bearing against the clutch release levers. Now, adjust

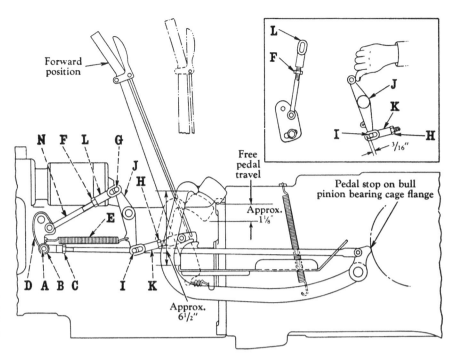

Fig. IH687A—Adjustment points for series 350U and 350DU engine and torque amplifier linkage.

clevis (K) to provide a space of $\frac{3}{16}$-inch between the inserted pin (I) and the forward end of the elongated hole in clevis (K). Tighten lock nut (H) and reinstall spring (E). Adjust the length of rod (N) with clevis (L) so that rod (N) is the shortest possible length that will not change the posi-

tion of levers (D & J) when pin (G) is inserted.

157. SERIES 300-350-350D-400-400D-W400-W400D-450-450D-W450-W450D. Refer to Fig. IH688. On series W400, W400D, W450 and W450D, remove battery and battery box. On all models, remove spring (6), loosen locknut (1) and remove clevis pin (3). Loosen locknut (9) and remove clevis pin (11). Turn clevis (10) until clutch pedal free travel (N) is as follows:

Series 300 ....$1\frac{1}{4}$ inches
Series 350-350D ....$1\frac{3}{4}$ inches
Series 400-400D ....$\frac{7}{8}$-inch
Series W400-W400D ....$\frac{11}{16}$-inch
Series 450-450D ....$1\frac{1}{4}$ inches
Series W450-W450D ....$\frac{15}{16}$-inch
Dimension (N) is measured horizontally from point of contact of clutch pedal lever and rear frame cover. After adjustment is complete, tighten lock nut (9). On models so equipped loosen lock nut (13) and turn the adjusting set screw (12) until the pedal full travel (M) is as follows:

Series 300 ....$3\frac{3}{4}$ inches
Series 400 ....4 inches
Series W400 ....$2\frac{13}{16}$ inches
Dimension (M) is measured horizontally from point of contact of clutch pedal lever and rear frame cover. When adjustment is complete, tighten lock nut (13).

After the engine clutch pedal linkage is properly adjusted, place the torque amplifier control lever in the forward position as shown and proceed to adjust the torque amplifier clutch linkage as follows:

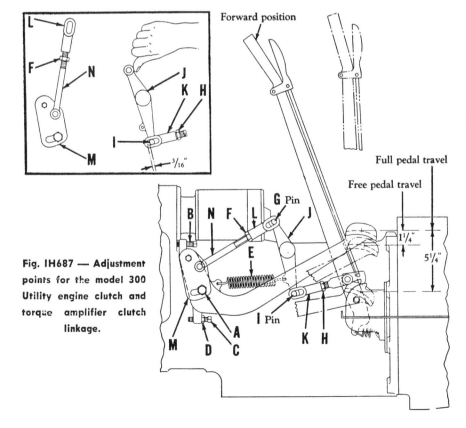

Fig. IH687 — Adjustment points for the model 300 Utility engine clutch and torque amplifier clutch linkage.

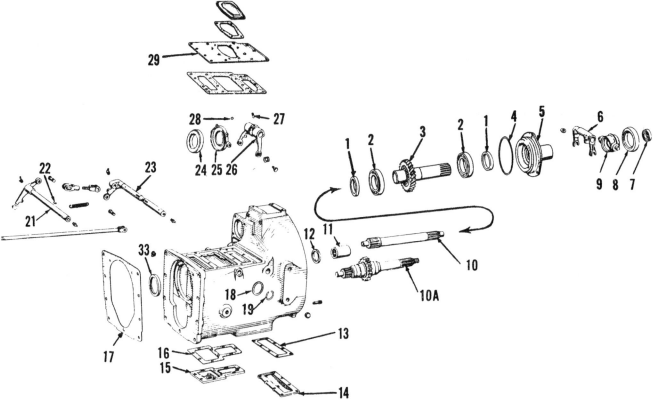

Fig. IH688—Adjusting points for the engine and torque amplifier clutch linkage. This illustration is typical of all series except the Utility models.

Loosen lock nut (5), remove clevis pin (7) and turn lever (4) counter-clockwise as far as possible without forcing. See inset. This places the TA clutch release bearing against the clutch release levers. Now, adjust clevis (8) to provide a space (P) of $\frac{3}{16}$-inch between the inserted pin (7) and the forward end of the elongated hole in clevis (8). Tighten locknut (5) and reinstall spring (6). Adjust the length of rod (14) with clevis (2) so that rod (14) is the shortest possible length that will not change the position of levers (4 and 15) when pin (3) is inserted.

159. R&R AND OVERHAUL. To remove the torque amplifier clutch cover assembly and lined plate, first detach (split) engine from clutch housing as outlined in paragraphs 144, 145 or 146 and proceed as follows:

On series 300, 350, 350D, 400, 400D, 450 and 450D, remove tool box. On series 300, 350, 350D, 400, 400D, W400, W400D, 450, 450D, W450 and W450D, raise belt pulley unit and remove the

Fig. IH690—Clutch housing and associated parts used on series 400 and 400D tractors. For the purposes of this illustration, other models are similar. Item (15) is a seasonal disconnect cover for models with independent power take-off.

1. Oil seal
2. Bearing
3. Independent pto drive gear
4. "O" ring seal
5. Pto drive gear bearing retainer
6. Engine clutch release fork
7. Clutch shaft pilot bearing
8. Clutch release bearing
9. Clutch release sleeve
10. Clutch shaft used on models with independent power take-off
10A. Clutch shaft and pto drive gear used on models without independent power take-off
11. Clutch shaft sleeve
12. Oil seal
13. Gasket
14. Cover
15. Seasonal disconnect cover
16. Gasket
17. Gasket
18. Washer
19. Snap ring
21. Torque amplifier clutch release shaft
22. Woodruff key
23. Engine clutch release shaft
24. Torque amplifier clutch release bearing
25. Release sleeve
26. Clutch release fork
27. Lock screw
28. Release sleeve spring
29. Clutch housing top cover
33. Oil seal

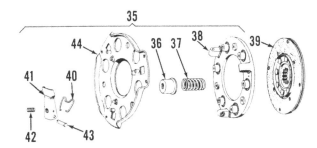

Fig. IH691—Exploded view of the torque amplifier clutch.

36. Spring cup
38. Pressure plate
39. Driven disc
40. Lever spring
41. Release lever
42. Adjusting screw
43. Lever pin

clutch housing top cover. On series 300U, 350U and 350DU, raise steering gear and fuel tank assembly and remove the clutch housing top cover. Disconnect linkage from the engine clutch release shaft (23 — Fig. IH690). loosen the cap screws retaining fork (6) to shaft and bump shaft toward side of housing until Woodruff keys are exposed. Extract Woodruff keys and withdraw the release shaft and the engine clutch release bearing and carrier (8 & 9). Unbolt and remove bearing cage (5) from clutch housing. Withdraw the independent power take-off drive shaft (3) and clutch shaft (10). Note: A tapped hole is provided in the end of the clutch shaft to aid in removal. On models without independent power take-off, the engine clutch shaft and pto drive gear are an integral unit as shown at (10A). Disconnect linkage from the TA clutch release shaft, remove lock screw (27) then remove snap ring (19) from right end of shaft. Withdraw shaft (21), fork (26) and release bearing and carrier (24 and 25). Use three $\frac{7}{16}$"-18 by $\frac{7}{16}$" cap screws and plain washers and screw them into the tapped holes provided in the pressure plate to keep the assembly under compression; then, unbolt the TA clutch cover assembly (35—Fig. IH691) from carrier and withdraw the clutch cover assembly and lined plate.

Fig. IH693—Using a special OTC socket to remove the torque amplifier clutch carrier retaining nut.

Fig. IH694—When adjusting the release lever height (L) of 1 5/8 inches, the back plate to pressure plate measurement (K) of 19/32 inch must be maintained.

Note: If carrier (47—Fig. IH692) is damaged, bend tang of locking washer (46) out of notch in nut (45), remove nut as shown in Fig. IH693 and bump carrier from splines of the primary sun gear.

Examine the driven plate for being warped, loose or worn linings, worn hub splines and/or loose hub rivets. Disassemble the clutch cover assembly and examine all parts for being excessively worn. The six pressure springs should have a free length of 2 inches and should require 161 lbs. to compress them to a height of 1¼ inches. Renew pressure plate if it is grooved or cracked. Renew back plate (44—Fig. IH691) if it is worn around the drive lug windows.

When reassembling, adjust the release levers to the following specifications. With a back plate to pressure plate measurement (K—Fig. IH694) of $\frac{19}{32}$-inch, the release lever height (L) from friction face of pressure plate to release bearing contacting surface of release levers is 1⅝ inches.

When reassembling, observe the clutch carrier and back plate for balance marks which are indicated by an arrow and white paint. If the balance marks are found on both parts, they should be assembled with the marks as close together as possible. If no marks are found, or if only one part is marked, the clutch balance can be disregarded. Install the remaining parts by reversing the removal procedure.

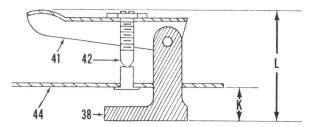

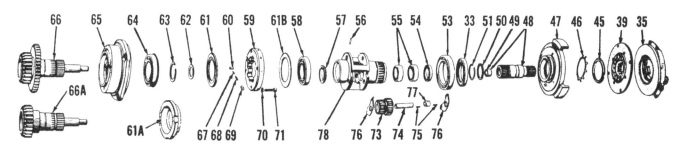

Fig. IH692—Exploded view of the torque amplifier unit. Planetary gears (73) are available in sets only. Item (61) is used on the 400, 400D, W400, W400D, 450, 450D, W450 and W450D series only and item (61A) is used on other series.

33. Oil seal	51. Snap ring	60. Clutch roller	66. Transmission drive shaft and secondary sun gear (except Utility)
35. Clutch cover assembly	53. Bearing	61. Rear thrust washer	
39. Driven plate	54. Oil seal	61A. Rear thrust washer	
45. Nut	55. Needle bearings	61B. Front thrust washer	66A. Transmission drive shaft and secondary sun gear (Utility)
46. Locking washer	56. Roll pins	62. Thrust washer	
47. Clutch carrier	57. Spacer	63. Snap ring	67. Pin
48. Primary sun gear	58. Bearing	64. Bearing	68. Spring
49. Needle bearing	59. Over-running clutch ramp	65. Transmission drive shaft bearing cage	69. Plug
50. Thrust washer			70. Retainer ring
			71. Allen head screw
			73. Planetary gears
			74. Gear shaft
			75. Needle bearing
			76. Thrust plate
			77. Bearing spacer
			78. Planet carrier

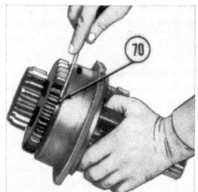

Fig. IH695—Removing retaining ring (70) from the Allen head over-running clutch cap screws.

Fig. IH696—Removing the Allen head over-running clutch cap screws.

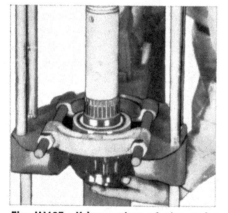

Fig. IH697—Using a piece of pipe and a press to remove front bearing from the planet carrier.

Fig. IH699—OTC tool number ED-3251. Press pilot bearing in the primary sun gear until surface (X) is even with rear edge of the primary sun gear.

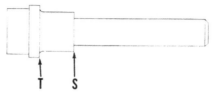

Fig. IH700—OTC tool number ED-3250 is used to install needle bearings in the planet carrier. Refer to text.

PLANET GEARS, SUN GEARS & OVER-RUNNING CLUTCH

All Models So Equipped

160. **R&R AND OVERHAUL.** To overhaul the torque amplifier gear set and over-running clutch, first remove the TA clutch cover assembly, lined plate and clutch carrier as outlined in paragraph 159 and proceed as follows: Remove all other parts attached to clutch housing. Remove the independent power take-off seasonal disconnect cover (15—Fig. IH690). Support rear half of tractor under rear frame and attach a chain hoist around clutch housing. Remove top bolt on each side of clutch housing and install aligning dowels (OTC tool No. ED-3271). Remove remaining bolts retaining clutch housing to transmission case and separate the units. Note: Lower center bolt connecting clutch housing to main frame is accessible through the seasonal disconnect opening.

Unbolt the transmission drive shaft bearing cage from clutch housing and withdraw the complete TA unit. Re-

move the small retainer ring (70—Fig. IH692 & IH695) from each of the four Allen head over-running clutch cap screws (71—Fig. IH692). Clamp the complete unit in a soft jawed vise and remove the four cap screws as shown in Fig. IH696. Note: A cutout is provided in the planet carrier for this purpose. Separate the transmission drive shaft and bearing cage assembly from planet carrier (78—Fig. IH692). Remove snap ring (63) from front of transmission drive shaft and press the transmission drive shaft rearward out of bearing and cage. Bearing (64) can be inspected and/or renewed at this time. Inspect ramp (59), springs (68) and rollers of over running clutch. Renew damaged parts. Using OTC bearing puller attachment 952-A or equivalent, press front and rear bearings (53 & 58) from the planet carrier. It is important, when removing the front bearing to use a piece of pipe and press against the planet carrier and **not** against the primary sun gear shaft. Refer to Fig. IH-697. Using a small punch and hammer, drive out the Esna roll pins retaining the planetary gear shafts in the planet carrier. Refer to Fig. IH698. Using OTC dummy shaft No. ED-3259, push out the planet gear shafts and lift gears with rollers and dummy

Fig. IH702—Chassis lubricant facilitates installation of needle bearings in the planetary gears.

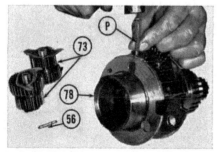

Fig. IH698—Using a punch and hammer to drive out the Esna roll pins which retain the planet gear shafts in the planet carrier.

P. Punch   73. Planet gears
56. Roll pin   78. Planet carrier

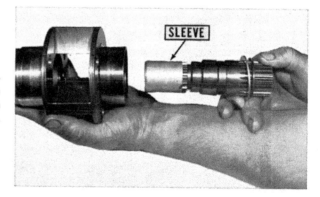

Fig. IH701—The seal protector sleeve is used when installing the primary sun gear in the planet carrier.

shaft out of the planet carrier. Be careful not to lose or damage the thrust plates (76—Fig. IH692) as they are withdrawn. After the three compound planetary gears are removed, the primary sun gear and shaft can be withdrawn from the planet carrier.

Inspect splines, oil seal surface, bearing areas, pilot bearing and sun gear teeth of the primary sun gear and shaft for excessive wear or damage. If only the pilot bearing (49) is damaged, renew the bearing. If any

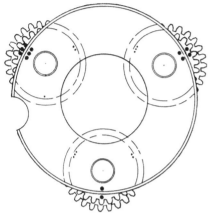

Fig. IH703—When planetary gears are installed properly, punch marks on gears and planet carrier will be in register.

other damage is found, renew the complete unit which includes an installed pilot bearing. Note: When installing a new pilot bearing, use OTC tool No. ED-3251 (shown in Fig. IH-699) and press the bearing in until surface (X) is even with rear edge of primary sun gear.

Inspect the planet carrier for rough oil seal surface, worn over-running clutch roller surface and elongated planet gear shaft holes.

Inspect the primary sun gear shaft needle bearings (55—Fig. IH692) and the shaft oil seal (54). If bearings and/or seal are damaged, and planet gear carrier is O. K., drive out the faulty parts with a brass drift and install new bearings using OTC driving collar ED-3250 (shown in Fig. IH-700). Press rear bearing in from front until surface (S) is even with front of planet carrier. Press the front needle bearing in from front until surface (T) is even with front of planet carrier. Install oil seal (54—Fig. IH-692) with lip toward rear until front of oil seal is even with front of planet carrier.

Install snap ring (51) on the primary sun gear shaft and thrust washer (50) immediately ahead of the snap ring. Using OTC oil seal protector sleeve

No. ED-3245, install the primary sun gear and shaft in the planet carrier as shown in Fig. IH701.

Inspect teeth of planet gears for wear or other damage. If any one of the three gears is damaged, renew all three gears which are available in a matched set only. These planetary gears are manufactured in matched sets so the gears will have an equal amount of backlash when installed and no one gear will carry more than its share of the load. Note: The International Harvester Co. specifies that when the planet gears are removed, the planet gear shafts and needle rollers should always be renewed. The planet gear shafts are available in sets of three and the needle rollers are available in sets of 138.

Using chassis lubricant and OTC dummy shaft ED-3259, install twenty-three new needle bearings in one end of a planet gear. Slide dummy shaft into bearings and install bearing spacer (77—Fig. IH692). With the aid of chassis lubricant, install twenty-three new needle bearings in the other end of the planet gear and slide the dummy shaft completely into the gear, thereby holding the needle bearings and spacer in the proper position.

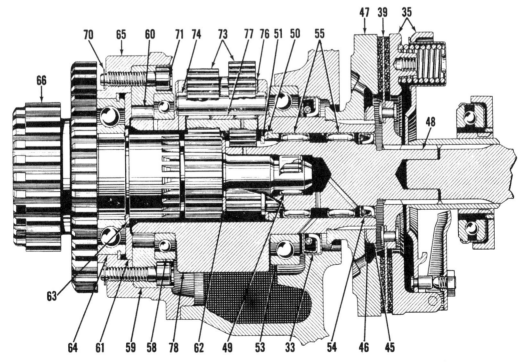

Fig. IH704—Typical sectional view of assembled torque amplifier unit. Bearings in the unit are non-adjustable.

33. Oil seal	49. Roller bearing	59. Over-running clutch ramp	65. Transmission drive shaft bearing cage	71. Allen head screw
35. Clutch cover assembly	50. Thrust washer	60. Clutch roller	66. Transmission drive shaft and secondary sun gear	73. Planetary gear
39. Driven plate	51. Snap ring	61. Thrust washer		74. Gear shaft
45. Nut	53. Bearing	62. Thrust washer		76. Thrust plate
46. Locking washer	54. Oil seal	63. Snap ring	70. Retainer ring	77. Bearing spacer
47. Clutch carrier	55. Needle bearing	64. Bearing		78. Planet carrier
48. Primary sun gear	58. Bearing			

Assemble one thrust plate (76) to each end of the planet gear and install planet gear, dummy shaft and thrust plates assembly in the planet carrier. Using one of the three new planet gear shafts, push out the dummy shaft and install the Esna roll pin securing planet gear shaft to the planet carrier. Refer to Fig. IH702. Observe the rear face of the planet carrier at each planet gear location where punched timing marks will be found. One location has one punch mark, another location has two punch marks and the other location has three. Turn the primary sun gear shaft until the timing punch mark (or marks) on the rear face of the installed planet gear are in register with the same mark (or marks) on the planet carrier. Now assemble the other planet gears, needle bearings, spacers, thrust plates and dummy shaft and install them so that timing marks are in register. When all three planet gears are installed properly, the single punch mark on planet carrier will be in register with single punch mark on one of the planet gears, double punch mark on carrier will register with double punch marks on one of the gears and triple punch marks on carrier will be in register with triple punch marks on one of the gears as shown in Fig. IH703.

Install thrust washer (61B—Fig. IH-692) on planet carrier. Assemble the pins (67—Fig. IH705), springs (68) and rubber plugs (69) into the over-running clutch ramp and place ramp on the planet carrier. Using a small screw driver, push pins (67) back and drop rollers (60) in place as shown. Install bearing (64—Fig. IH704) with snap ring in the rear bearing cage (65), press the transmission drive shaft (66) into position and install snap ring (63).

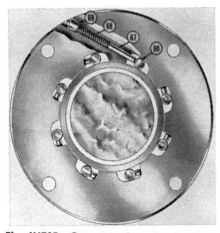

Fig. IH705—Cut-away view shows proper installation of over-running clutch rollers (60), pins (67), springs (68) and plugs (69) in ramp.

Place the planet carrier on the bench with rear end up and lay thrust washer (62) on the primary sun gear. Place the over-running clutch thrust washer (61 or 61A) on the ramp so that polished surface of thrust washer will contact rollers. Install the assembled transmission drive shaft and bearing cage and secure in position with the four Allen head cap screws (71). After the cap screws are tightened to a torque of 40 ft.-lbs., install

the small retainer rings (70).

Inspect the large oil seal (33) in the clutch housing and renew if damaged. Lip of seal goes toward rear of tractor.

Using OTC oil seal protector sleeve No. ED-3253 over splines of planet carrier, insert the assembled TA unit and tighten the transmission drive shaft bearing cage cap screws securely. Assemble the remaining parts by reversing the disassembly procedure.

TRANSMISSION

The transmission, differential and final drive gears are all contained in the same case which is called the rear frame. A wall in the case separates the bull gear and differential compartment from the transmission gear set. Shifter rails and forks are mounted on underside of the rear frame (transmission) cover.

All bearings in the transmission are of the non-adjustable ball type.

TOP COVER

Series 300U-350U-350DU

161. To remove the transmission top cover, drain the hydraulic system and on models so equipped, disconnect the power steering oil lines. Remove the Hydra-Touch manifold connecting the pump to the reservoir. Remove the seat and tool box. Unbolt and remove the transmission top cover and hydraulic reservoir assembly from tractor.

Series 300-350-350D-400-400D-W400-W400D-450-450D-W450-W450D

162. There are several procedures which can be followed when removing the transmission top cover. The I&T recommended procedure, which is probably the quickest, is as follows:

Detach (split) transmission housing from the clutch housing as outlined in paragraph 164 for series 300, 350, 350D, 400, 400D, 450 and 450D, paragraph 165 for series W400, W400D, W450 and W450D; then, unbolt and remove the top cover from the transmission case.

TRACTOR SPLIT

Series 300U-350U-350DU

163. To detach (split) the transmission from the clutch housing, first drain the transmission, hydraulic system and torque amplifier housing; then disconnect the hydraulic manifold, power steering lines, tail light

wires and clutch rod. Remove the TA control lever assembly, both platforms and the pto seasonal disconnect cover from bottom of clutch housing. Disconnect the battery ground strap. Support both halves of tractor, then unbolt and separate the tractor halves.

Series 300-350-350D-400-400D-450-450D

164. To detach (split) the transmission from the clutch housing, drain the transmission and torque amplifier and remove hood. Shut-off the fuel and remove the fuel line. Loosen the fuel tank support bolts and disconnect the hydraulic manifold at pump. Remove the battery cable cover and the cap screws retaining the hydraulic reservoir to transmission top cover. Unbolt the hydraulic check valve unit from battery box and remove the belt pulley control rod. Disconnect the throttle rod and steering shaft universal joint and pull the steering shaft rearward. Disconnect the battery cable from starter switch and unbolt the hood support bracket from right side of tractor. Attach hoist to steering shaft, pull the hydraulic manifold outward to clear pump and raise the fuel tank and hydraulic reservoir approximately five inches. Remove the belt pulley unit.

Disconnect the clutch and torque amplifier control rods and remove the pto seasonal disconnect cover from bottom of clutch housing. Support both halves of tractor separately, unbolt clutch housing from transmission and roll transmission half of tractor rearward.

Series W400-W400D-W450-W450D

165. To detach (split) transmission from clutch housing, drain the transmission and torque amplifier housings and remove the hood. Disconnect the

engine speed control linkage and Hydra-Touch control handle bracket from steering column. Disconnect the drag link from the steering (Pitman) arm. Remove the steering gear housing shields; then unbolt and remove the steering gear unit from transmission cover. Remove the belt pulley unit, batteries and battery box. Unbolt the convenience outlet and hydraulic hoses from platform. Remove the cap screws retaining the hydraulic reservoir to transmission top cover.

Attach hoist to the hydraulic reservoir and raise the reservoir approximately one inch. Disconnect the clutch and torque amplifier rods, and remove the pto seasonal disconnect cover from bottom of clutch housing. Support both halves of tractor separately, unbolt clutch housing from transmission and roll transmission half of tractor rearward.

OVERHAUL
All Models

Data on overhauling the various transmission components are outlined in the following paragraphs. In general, the following paragraphs apply to all models; however, in some cases, the overhaul procedures differ on models equipped with and without torque amplifier as well as those equipped

with and without independent power take-off. Where these differences are encountered, they will be mentioned.

175. **SHIFTER RAILS AND FORKS.** Shifter rails and forks are retained to bottom side of the rear frame (transmission) cover and are accessible for overhaul after removing the cover as outlined in paragraph 161 or 162. The overhaul procedure is conventional and evident after an examination of the unit.

176. **TRANSMISSION DRIVING SHAFT.** On models with torque amplifier, the transmission driving gear and shaft (66—Fig. IH706) is integral with the torque amplifier secondary sun gear and is normally serviced in conjunction with overhauling the torque amplifier unit as outlined in paragraph 160.

On models without torque amplifier, the transmission driving shaft is considered a part of the transmission. To remove the drive shaft, it is first necessary to detach (split) the transmission housing from the clutch housing as outlined in paragraph 163, 164 or 165.

With the transmission detached from the clutch housing, the procedure for removing and overhauling the drive shaft is evident after an exam-

ination of the unit and reference to Figs. IH707, 711 and 716.

177. **MAINSHAFT PILOT BEARING.** To remove the mainshaft pilot bearing (9—Fig. IH707), detach (split) the transmission housing from the clutch housing as outlined in paragraph 163, 164 or 165. Remove the bearing retaining plate from front end of mainshaft and using OTC bearing puller or equivalent as shown in Fig. IH708, remove the pilot bearing.

178. **MAINSHAFT (SLIDING GEAR OR BEVEL PINION SHAFT).** To remove the transmission main shaft, first detach (split) transmission from clutch housing as in paragraph 163, 164 or 165 and remove the transmission top cover.

Remove the three cap screws retaining the mainshaft rear bearing retainer (5—Fig. IH707) to the main case dividing wall, move the mainshaft assembly forward and withdraw the unit, rear end first, as shown in Fig. IH709.

Remove the mainshaft pilot bearing (9—Fig. IH707) and slide gears from shaft. Remove snap ring (4) and press or pull rear bearing and retainer from shaft. When reassembling, install bearing (3) so that ball loading grooves are toward front or away

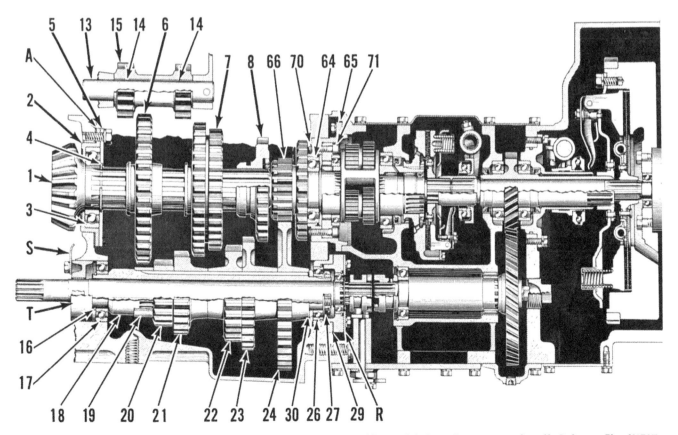

Fig. IH706—Sectional view of a typical transmission with torque amplifier and independent power take off. Refer to Fig. IH707.

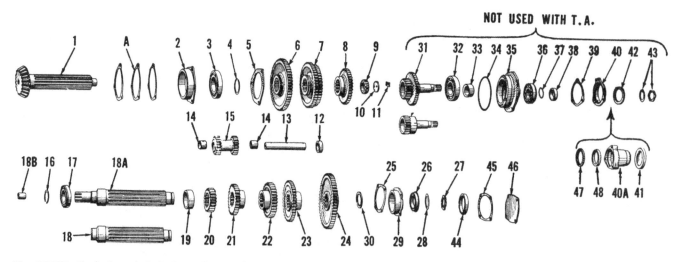

Fig. IH707—Typical exploded view of transmission shafts, gears and associated parts. All bearings are of the non-adjustable ball type. Main drive bevel pinion position is controlled by shims (A). Drive shaft used on utility models is shown below item 31.

1. Mainshaft and bevel pinion	14. Reverse idler bushing	20. First speed driving gear	30. Spacer
2. Bearing cage	15. Reverse idler gear	21. Second speed driving gear	31. Transmission drive shaft (not utility)
3. Rear bearing	16. Snap ring	22. Third speed driving gear	32. Ball bearing
5. Bearing retainer	17. Bearing	23. Fourth speed driving gear	33. Spacer
6. First and reverse sliding gear	18. Countershaft with IPTO	24. Constant mesh gear	34. Seal ring
7. Second and third sliding gear	18A. Countershaft with no IPTO	26. Bearing	35. Bearing cage
8. Fourth and fifth sliding gear	18B. Bushing with no IPTO	27. Nut	36. Ball bearing
13. Reverse idler shaft	19. Spacer	29. Bearing cage	37. Seal ring
			38. Spacer
			39. Gasket
			40. Bearing retainer

40A. Retainer, used with IPTO
41. Oil seal
43. Nut and lock washer
44. Spacer
45. Gasket
46. Cap
47. Spacer, 300, 300U, 350, 350D, 350U and 350DU
48. Seal

from the bevel pinion gear. Use Fig. IH706, 711 or 716 as a guide when installing the sliding gears and if the same mainshaft is installed, be sure to use the same shims (A) as were removed. If a new mainshaft is being installed, use the same shims (A) as a starting point, but be sure to check and adjust if necessary the main drive bevel gear mesh position as outlined in the main drive bevel gear section.

179. COUNTERSHAFT. To remove the transmission countershaft, first remove the mainshaft as outlined in paragraph 178 and proceed as follows: On models with independent power take-off, remove the four cap screws retaining the independent power take-off extension shaft front bearing cage to main frame and withdraw the extension shaft, bearing cage and retainer as shown in Fig. IH710. Working in the bull gear compartment of the main frame, remove the independent power take-off coupling shaft and the extension shaft rear bearing carrier retainer strap (S—Fig. IH706).

On models with transmission driven power take-off, remove the pto shaft assembly and cap (46—Fig. IH707).

On all models, remove the nut from forward end of countershaft and while bucking up the countershaft gears, bump the countershaft rearward until free from front bearing. Remove the countershaft front bearing and cage.

Withdraw the countershaft from rear and remove gears from above.

The rear bearing can be removed from countershaft after removing snap ring (16—Fig. IH707). When reassembling, use Fig. IH706, 711 or 716 as a guide and make certain that beveled edge of constant mesh gear spacer (30) is facing toward front of tractor.

180. REVERSE IDLER. With the countershaft removed as outlined in paragraph 179, the procedure for removing the reverse idler is evident. Bushings (14—Fig. IH707) are renewable and should be reamed after installation, if necessary, to provide a recommended clearance of 0.003-0.005 for the idler gear shaft.

Fig. IH708—Using OTC puller to remove pilot bearing from front of transmission main shaft.

Fig. IH709 — Removing the transmission mainshaft assembly.

Fig. IH710 — Removing the independent power take-off extension shaft, front bearing and cage.

MAIN DRIVE BEVEL GEARS AND DIFFERENTIAL

ADJUST BEVEL GEARS
All Models

185. The tooth contact (mesh pattern) and backlash of the main drive bevel gears is controlled by shims on all models. Tooth contact (mesh pattern) and backlash should be checked and adjusted, if necessary, whenever the transmission is being overhauled, and always when a new pinion or ring gear, or both, are installed.

186. **MESH AND BACKLASH.** The first step in adjusting a new set of bevel gears is to arrange shims (B—Fig. IH720) to provide the desired backlash of 0.008-0.012 between the main drive bevel pinion and ring gear.

The next step is to arrange shims (A — Fig. IH706, 711 or 716) located between the transmission mainshaft rear bearing cage and the rear frame to provide the proper tooth contact (mesh position) of the bevel gears.

Paint the pinion teeth with Prussian blue or red lead and rotate the ring in normal direction of rotation and observe the contact pattern on the tooth surfaces.

The area of **heaviest contact** will be indicated by the coating being **removed** at such points. On the actual pinion, the tooth contact areas shown in black on the illustrations will be bright; that is, there will be no blue or red coating on them.

The desired condition is indicated in Fig. IH713, which shows that the paint has been removed from the toe end of the teeth over the distance A to B as shown.

When the heavy contact is concentrated high on the toe at A as shown in Fig. IH714, the pinion should be moved toward front of tractor by adding a shim behind the pinion bearing cage.

When the heavy contact is concentrated low on the toe of the pinion tooth as in Fig. IH715, remove a shim

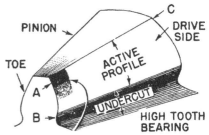

Fig. IH714—High tooth contact at "A" (no load) indicates that the pinion has been set too far in.

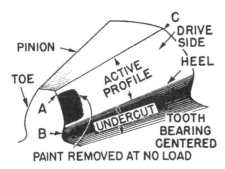

Fig. IH713—Tooth contact should be centered between "A" and "B" (no load). When a heavy load is placed on the gears, the tooth contact will extend from the toe almost to the heel of the tooth.

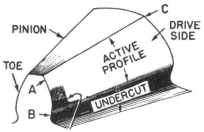

Fig. IH715—Low tooth contact at "B" (no load) indicates that the pinion has been set too far out.

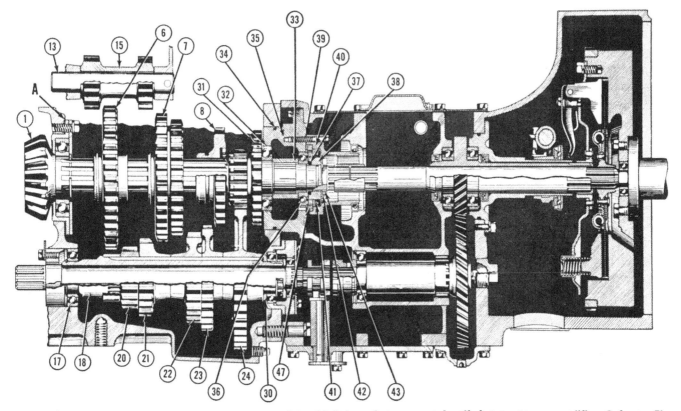

Fig. IH711—Typical transmission sectional view on models with independent power take-off, but no torque amplifier. Refer to Fig. IH707 for legend.

from the pinion bearing cage.

After obtaining desired tooth contact, recheck the backlash and if not within the desired limits 0.008-0.012, adjust by transferring a shim or shims from behind one differential bearing cage to the other bearing cage until desired backlash is obtained.

Do not expect the contact pattern to extend much farther toward the heel of the pinion than shown in Fig. IH713. The teeth are purposely ground to produce a toe bearing or contact under zero or light load so that when the gear supports deflect and the teeth deform under heavy load conditions, the contact pattern will increase in area along the teeth toward the heel and thus automatically increase the load carrying capacity of the gear.

RENEW BEVEL GEARS

All Models

187. To renew the main drive bevel pinion, follow the procedure outlined in paragraph 178 for overhaul of the transmission main shaft. To renew the main drive bevel ring gear, follow the procedure outlined for overhaul of the differential assembly (paragraph 188).

DIFFERENTIAL AND CARRIER BEARINGS

All Models

Differential unit is of the four pinion type mounted back of a dividing wall in the rear frame (transmission case). Refer to Fig. IH720.

The differential case halves are held together by bolts which also retain the bevel ring gear. The bearings supporting the differential are of the non-adjustable ball type.

188. **R&R AND OVERHAUL.** To remove the differential and the main drive bevel ring gear assembly, first

remove the final drive bull gears as outlined in paragraph 190 and proceed as follows: Remove the brake housings, brake discs and the inner brake plates (1—Fig. IH720 or 721). Remove the bull pinion shaft bearing cages (6 and 20) and lift the differential and

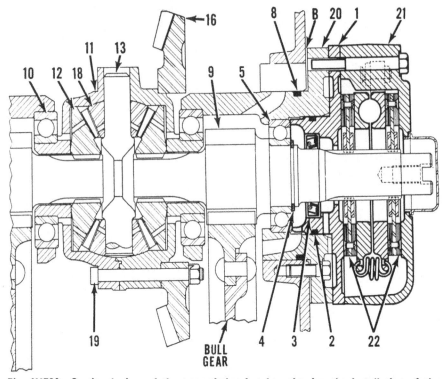

Fig. IH720—Sectional view of the transmission housing, showing the installation of the differential, bull pinions and disc brakes. All bearings in the final drive are of the non-adjustable ball type. Refer to Fig. IH721 for legend.

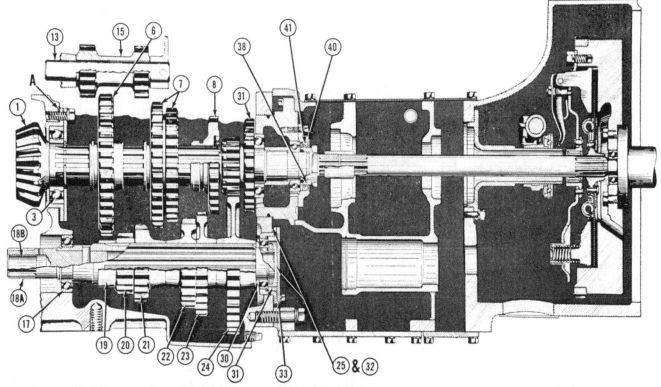

Fig. IH716—Typical transmission sectional view on models with notorque amplifier and no independent power take-off. Refer to Fig. IH707 for legend.

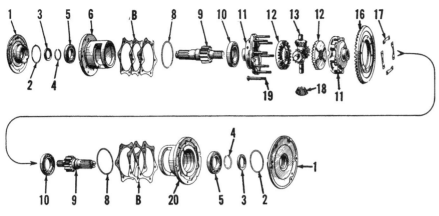

Fig. IH721—Exploded view of the differential, bull pinions and associated parts. Shims (B) control backlash of the main drive bevel gears. Either bull pinion can be removed without removing the respective bull gear.

B. Shims
1. Bearing retainer (inner brake plate)
2. Seal ring
3. Bull pinion shaft oil seal
4. Snap ring
5. Ball bearing
6. Left bull pinion shaft bearing cage
8. Seal ring
9. Bull pinion
10. Differential carrier bearing
11. Differential case half
12. Differential side gear
13. Spider
16. Bevel ring gear
17. Lock plates
18. Pinions
19. Case bolts
20. Right bull pinion shaft bearing cage

bevel ring gear assembly from tractor. Save and do not mix shims (B). Note: The differential carrier bearings can be renewed at this time.

The procedure for overhauling the removed differential unit is evident after an examination of Fig. IH721.

When reinstalling the differential unit, assemble bull pinion shaft (9) to left hand bearing cage (6) and install the unit. Lower differential unit into rear frame and enter same over splines of bull pinion shaft. Install opposite (right) bearing cage and bull pinion unit.

The differential carrier bearings are non-adjustable, but shims (B) are provided to position the main drive bevel ring gear for proper backlash in relation to the bevel pinion. Desired backlash of 0.008-0.012 is obtained by removing a shim or shims (B) from one bearing cage and inserting them under the other bearing cage. Refer to paragraphs 185 and 186 for method of adjusting tooth contact (mesh pattern) of main drive bevel gears.

FINAL DRIVE
All Models

As treated in this section, the final drive will include the bull pinions and integral shafts, both bull gears and both rear wheel axle and sleeve assemblies.

189. **R&R BULL PINION (DIFFERENTIAL) SHAFTS.** To remove either bull pinion shaft (9—Fig. IH-720), remove the respective brake unit, remove the bull pinion shaft bearing retainer (1) (inner brake drum) and withdraw the bull pinion shaft and bearing from the bearing cage.

190. **R&R BULL GEAR.** To remove either bull gear, first remove the transmission top cover as in paragraph 161 or 162. Remove the rear wheel and cap screw (M—Fig. IH722 or 723) retaining bull gear to inner end of wheel axle or bull sprocket shaft. Remove cap screws retaining rear axle or bull sprocket housing or sleeve to transmission housing and withdraw

carrier and shaft as a unit from the rear frame. (On high clearance models withdraw carrier and one final drive unit as an assembly.) Bull gear can now be removed from rear frame (transmission housing).

191. **R&R WHEEL AXLE SHAFT.** To renew wheel axle shaft or bearings on all except high clearance models, first detach bull gear as described in paragraph 190. With the axle shaft and carrier unit off tractor, remove outer bearing retainer and remove shaft from carrier by bumping on its inner end. Refer to Fig. IH722.

191A. To remove wheel axle shaft or sprocket on high clearance models, proceed as follows: After draining the lubricant, remove housing pan (31—Fig. IH723) and the connecting link from drive chain (25). It is not necessary to remove the drive chain. If removal of drive chain is desired, fasten a piece of flexible wire to one end and thread chain off of sprocket (20). The wire is used to help install and thread the chain over the top sprocket.

Remove inner and outer wheel axle bearing retainers (27 and 32) and inner bearing cap screw (33). Bump wheel axle shaft (28) on inner end and out of sprocket while bucking up the sprocket (44).

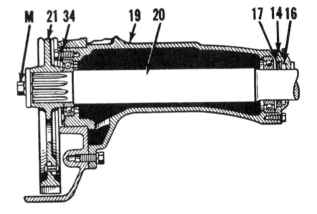

Fig. IH722 — Sectional view showing a typical rear axle and housing used on all except high clearance models. Bull gear (21) is retained to inner end of axle shaft by cap screw (M).

M. Cap screw
14. Bearing retainer
16. Felt seal
17. Oil seal
19. Axle housing
20. Axle shaft
21. Bull gear
34. Bearing retainer

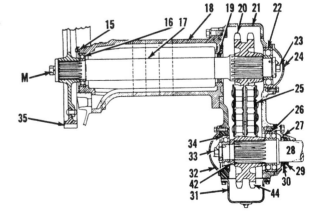

Fig. IH723 — Sectional view of the final drive unit used on high clearance models.
M. Cap screw
15. Bearing retainer
16. Bearing
17. Sprocket shaft
18. Housing carrier
19. Oil seal
20. Drive sprocket
21. Rear axle housing
22. Bearing
23. Cap screw
24. Bearing cap
25. Drive chain
26. Bearing
27. Bearing retainer
28. Rear axle
29. Felt seal
30. Oil seal
31. Housing pan
32. Bearing cap
33. Cap screw
34. Bearing
35. Bull gear

When reinstalling axle shaft, install spacer (42) between inner bearing (34) and sprocket hub with wide face of spacer nearest the sprocket hub.

192. **R&R SPROCKET (UPPER) SHAFT.** To remove the upper sprocket shaft (17—Fig. IH723) on high clearance models proceed as follows: Remove bull gear as outlined in paragraph 190. Remove drive chain as outlined in first part of paragraph 191A. Remove outer bearing cap (24) and cap screw (23) retaining bearing (22) to shaft. Remove cap screws retaining sprocket shaft carrier (18) to transmission housing and remove carrier (18). Bump sprocket shaft out of sprocket and withdraw the sprocket.

BRAKES

All Models

200. Brakes are of the double disc, self energizing type, which are splined to the outer ends of the bull pinion shaft. The molded linings are riveted to the brake discs. Procedure for removing the lined discs is evident after an examination of the unit. CAUTION: Always be sure there is plenty of clearance between the actuating disc pivot lugs and brake housing. It may be necessary to grind the housing slightly as shown in Fig. IH724A, to obtain the desired $\frac{1}{16}$-inch clearance.

201. **ADJUSTMENT.** To adjust the brakes, loosen jam nut (5—Fig. IH724) and turn rod (1) either way as required to obtain the correct pedal free travel of 1⅛—1⅜ inches measured at top of rear frame cover. Brakes can be equalized by loosening the tight brake. When adjustment is completed, tighten jam nut (5). If for any reason the brake operating rod assembly has been disassembled, spring (2) should be adjusted to a pre-load dimension of $1\frac{1}{16}$ inch for series 300, 300U, 350, 350U, 350D and 350DU, $1\frac{3}{16}$ inch for series 400, 400D, W400, W400D, 450, 450D, W450 and W450D. Assemble component parts as shown and turn nut (3) until the correct dimension is obtained, and lock same by tightening jam nut (4).

Fig. 724A—There should be a minimum of 1/16-inch clearance between the actuating disc pivot lugs and inside of brake housing. It may be necessary on some tractors to grind the housings to obtain this clearance.

Fig. IH724—Cut-away view showing a typical disc type brake.

1. Adjusting rod
2. Spring
3. Nut
4. Nut
5. Nut
6. Housing
7. Lined disc
8. Actuating disc
9. Spring
10. Bearing retainer (inner brake drum)
11. Bearing quill
12. Ball.

BELT PULLEY

Series 300-350-350D-400-400D-W400-W400D-450-450D-W450-W450D

202. **REMOVE AND REINSTALL.** The procedure for removing the belt pulley unit on series 400, 400D, W400, W400D, 450, 450D, W450 and W450D is evident after an examination of the installation. To remove the belt pulley on series 300, 350 and 350D, proceed as follows:

Drain the hydraulic system and remove hood. Shut off the fuel and remove the fuel line. Loosen the fuel tank support bolts and disconnect the hydraulic manifold at pump. Remove the battery cable cover and the cap screws retaining the hydraulic reservoir to the transmission top cover. Unbolt the hydraulic check valve unit from battery box and remove the belt pulley control rod. Disconnect the throttle rod and steering shaft universal joint and pull the steering shaft rearward. Disconnect the battery cable from starter switch and unbolt the hood support bracket from right side of tractor. Attach hoist to steering shaft, pull the hydraulic manifold outward to clear pump and raise the fuel tank and hydraulic reservoir approximately five inches. Unbolt and remove pulley from tractor.

203. OVERHAUL. The procedure for disassembling and reassembling the belt pulley is evident after an examination of the unit and reference to Fig. IH725. Shims (15 and 24) control the mesh position and backlash of the bevel gears. Recommended bevel gear backlash is 0.008-0.010.

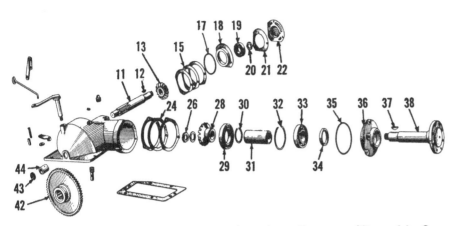

Fig. IH725—Exploded view of belt pulley unit used on all except utility models. Gear (42) meshes with spur gear on the transmission drive shaft.

11. Drive shaft	21. Gasket	33. Ball bearing
12. Woodruff key	22. Bearing cage cover	34. Oil seal
13. Bevel drive pinion	24. Shims	35. Seal ring
15. Shims	26. Nut	36. Bearing cage
17. Seal ring	28. Driven bevel gear	37. Woodruff key
18. Drive shaft	29. Ball bearing	38. Pulley shaft
bearing cage	30. Snap ring	42. Driving spur gear
19. Ball bearing	31. Spacer	43. Cup plug
20. Nut	32. Snap ring	44. Drive shaft bushing

POWER TAKE-OFF

All Models (Non-Continuous Type)

206. A cross-sectional view of the non-continuous (transmission driven) pto is shown in Fig. IH727. The forward end of the pto shaft (10) is supported in pilot bearing (21) in the end of the transmission countershaft. The procedure for renewing the transmission countershaft is outlined in paragraph 179. Lips of seal (16) should face inward.

All Models (Independent Type)

207. OVERHAUL. The occasion for overhauling the complete power take-off system will be infrequent. Usually, any failed or worn part will be so positioned that localized repairs can be accomplished. The subsequent paragraphs will be outlined on the basis of local repairs. If a complete overhaul is required, refer to the following paragraphs.

208. ADJUST REACTOR BANDS. To adjust the reactor bands, remove the adjusting screw cover, loosen lock nuts (N—Fig. IH728) and back off the adjusting screws (M) approximately four turns. Hold the operating lever so that punch marked dot on pawl is aligned with the punch dot on quadrant as shown at (X).

Turn the adjusting screws in (clockwise) until screws are reasonably tight and set the lock nuts finger tight. Move the operating lever back and forth several times; then, back-off the adjusting screws approximately one turn and tighten the lock nuts.

Using a $\frac{5}{16}$-inch rod approximately 8 inches long as a lever in hole of pto shaft, check to make certain that shaft is free to turn only when punched dots (X) on pawl and quadrant are aligned. The shaft should not turn with lever in any other position.

If shaft turns freely in any other position (than X) the adjusting screws are not evenly adjusted and readjustment is necessary.

If shaft is not free to turn in any position of the lever, turn the adjusting screws **out** (counter-clockwise) one half turn, lock and recheck.

Tighten locknuts when adjustment is complete.

209. RENEW REACTOR BANDS. To renew the reactor bands (34—Fig. IH729), remove the pto shaft guard and the band adjusting screw cover. Unbolt bearing retainer (47) from the housing cover and remove the retainer. Unlock the pto shaft nut (45),

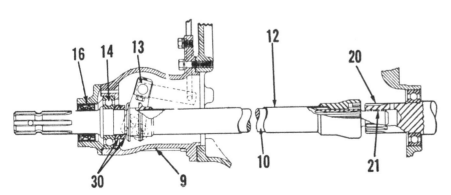

Fig. IH727—Sectional view of a typical non-continuous (transmission driven) power take-off shaft. The unit is driven by the splined rear end of the transmission countershaft (20).

9. Pto housing	16. Oil seal
10. Pto shaft	20. Transmission countershaft
12. Shifter coupling	21. Pilot bushing
13. Shifter	30. Double nut
14. Pto shaft outer bearing	

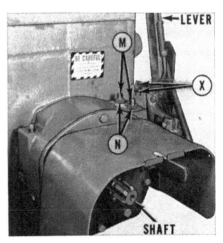

Fig. IH728—When adjusting the independent power take-off reactor bands, punched mark on pawl should be aligned with the punch mark on the quadrant.

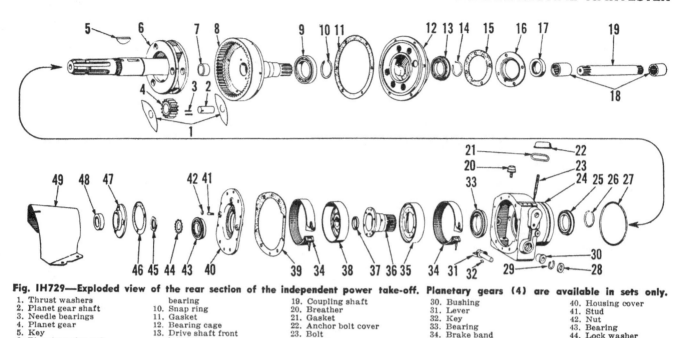

Fig. IH729—Exploded view of the rear section of the independent power take-off. Planetary gears (4) are available in sets only.

1. Thrust washers	bearing	19. Coupling shaft	30. Bushing	40. Housing cover
2. Planet gear shaft	10. Snap ring	20. Breather	31. Lever	41. Stud
3. Needle bearings	11. Gasket	21. Gasket	32. Key	42. Nut
4. Planet gear	12. Bearing cage	22. Anchor bolt cover	33. Bearing	43. Bearing
5. Key	13. Drive shaft front	23. Bolt	34. Brake band	44. Lock washer
6. Planet carrier and	bearing	24. Housing	35. Brake drum	45. Nut
pto shaft	14. Snap ring	25. Bearing	36. Sun gear	46. Gasket
7. Needle bearing	15. Gasket	26. Snap ring	37. Spacer	47. Bearing retainer
8. Ring gear and	16. Bearing retainer	27. Seal ring	38. Creeper drum	48. Oil seal
shaft	17. Oil seal	28. Oil seal	39. Gasket	49. Shaft guard
9. Drive shaft rear	18. Couplings	29. Snap ring		

place the operating lever in the forward position to prevent shaft from turning and remove nut (45). Remove the cap screws retaining the rear cover to the housing, place operating lever in the neutral position and turn the pto shaft to align threaded holes in the rear drum (38) with the unthreaded holes in the rear cover.

Using OTC puller ED-3262 or equivalent as shown in Fig. IH730, remove the rear drum and cover. Loosen the band adjusting screws and remove bands. Inspect bands for distortion, lining wear and for looseness of strut pins. When reassembling, adjust the bands as outlined in paragraph 208.

210. OVERHAUL PLANET GEARS, SUN GEARS AND SHAFTS. To overhaul the pto rear unit, first drain

transmission, then remove the reactor bands as outlined in paragraph 209. Unbolt and remove the unit from the tractor.

With the unit on bench, remove cap screws securing bearing cage (12—Fig. IH729 or IH733A) to housing and remove bearing cage with ring gear and shaft. Remove the front bearing retainer (16) and seal (17). Remove snap ring (14) and press the ring gear and shaft from bearing (13). The front bearing (13) can be removed from cage at this time. To remove the ring gear rear bearing (9), remove snap ring (10) and using a punch through the two holes in the ring gear hub, bump bearing from shaft. Inspect needle roller pilot bearing (7). If the bearing is damaged, it can be re-

moved, using a suitable puller as shown in Fig. IH732.

Withdraw the planet carrier and pto shaft from housing and mark the rear face of each planet gear so it can be installed in the same position. Using a punch as shown in Fig. IH733, remove the Esna roll pins which retain the planet gear shafts in the planet carrier and remove the shafts, gears, spacers and needle bearings. Remove snap ring (26—Fig. IH729 or

Fig. IH730—Using OTC puller to remove cover and rear drum from the independent power take-off rear unit.

Fig. IH732 — Removing the needle pilot bearing from ring gear.

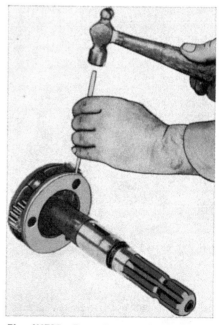

Fig. IH733—Removing the Esna roll pins which secure the planet gear shafts in the planet carrier.

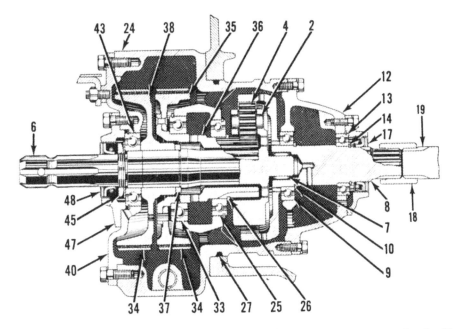

Fig. IH733A—Sectional view of the independent power take-off rear unit. Refer to legend under Fig. IH729.

IH733A) from sun gear and press sun gear (36) and drum (38) from housing. Bearing (25) can be removed from housing at this time. If rear bearing (33) is damaged, use a punch through holes in hub and drift the bearing from the sun gear as shown in Fig. IH734. Remove operating linkage from side of housing. To disassemble the spring retainer plug assembly, turn spring anchor block clockwise to relieve spring pressure and remove snap ring.

Caution: If the anchor bolt is broken or if spring tension cannot be relieved, use care when removing the snap ring.

Inspect all parts and renew any which are excessively worn. The planet gears are available only in sets as are the planet gear needle bearings and the gear shafts. Friction surface of drums should not be excessively worn.

When reassembling, reverse the disassembly procedure. OTC dummy shaft No. ED-3258-1 is used to assemble the needle bearings in the planet gears and to install the planet gears, with thrust plates to the planet carrier. With the planet gear, thrust plates and dummy shaft in position in the planet carrier, push the dummy shaft out with the new shaft. Secure the planet gear shafts to the planet carrier with the Esna roll pins.

When installation is complete, adjust the reactor bands as outlined in paragraph 208.

211. COUPLING SHAFT. To renew the pto coupling shaft (19—Fig. IH733A), remove the complete pto rear unit (Fig. IH735) and withdraw coupling shaft from rear frame.

212. EXTENSION SHAFT. To remove the pto extension shaft (Fig. IH-737), first detach (split) clutch housing from rear frame as outlined in paragraph 163, 164 or 165. Remove the seasonal disconnect coupling from front of shaft. Unbolt the extension shaft front bearing retainer and cage from transmission and withdraw the

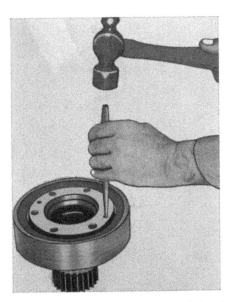

Fig. IH734—Using a punch through holes of sun gear to remove rear bearing.

Fig. IH735—Rear view of the tractor rear frame, showing the installation of the independent power take-off rear unit.

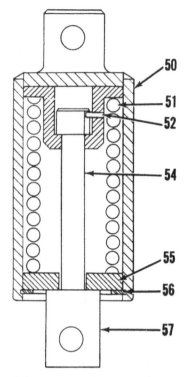

Fig. IH736—Sectional view of spring retainer plug assembly. Refer to caution in text before disassembling.

50. Spring sleeve   55. Retainer plate
52. Roll pin   56. Snap ring
54. Anchor bolt   57. Anchor block

extension shaft assembly as shown in Fig. IH738.

Remove the coupling shaft. Remove cap screw and strap which retains the extension shaft rear bushing carrier (Fig. IH737) in the rear frame and remove the bushing carrier. Bushing can be renewed if it is worn.

Reinstall the extension shaft by reversing the removal procedure.

213. DRIVEN SHAFT AND GEAR. Refer to Fig. IH737. To remove the pto driven shaft and gear, first remove the complete clutch and torque amplifier housing and proceed as follows:

Remove the driven gear cover from bottom of clutch housing and the large pipe plug which is located directly in front of the driven shaft. Remove cap screw and washer retaining driven gear to shaft and remove snap ring from behind the driven shaft rear bearing. Using a brass drift, bump driven shaft rearward out of clutch housing and withdraw the driven gear. If the driven shaft front needle bearing is damaged, it can be renewed at this time. The needle bearing race on shaft can also be renewed if damaged. Remove the race

Fig. IH738 — Removing the independent power take-off extension shaft from transmission housing.

with a brass drift to avoid damaging the shaft.

When reassembling, reverse the disassembly procedure and use sealing compound around pipe plug in front of driven gear.

214. DRIVE SHAFT. Refer to Fig. IH737. To remove the driving shaft and integral gear, first detach (split) engine from clutch housing as outlined in paragraph 144, 145 or 146 and remove the engine clutch release bearing and shaft. Unbolt the drive shaft front bearing cage and withdraw the drive shaft and bearing cage from clutch housing. The need and procedure for further disassembly is evident.

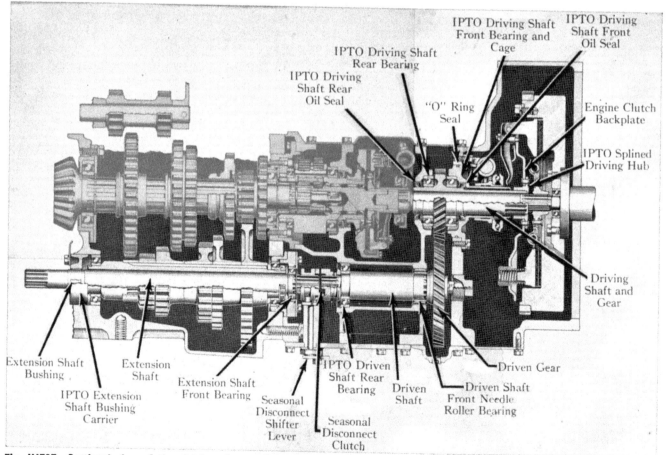

Fig. IH737—Sectional view of the clutch and transmission housing, showing the installation of the independent power take-off shafts and gears.

HYDRAULIC LIFT SYSTEM

NOTE: The maintenance of absolute cleanliness of all parts is of utmost importance in the operation and servicing of the hydraulic system. Of equal importance is the avoidance of nicks or burrs on any of the working parts.

LUBRICATION
All Models

220. It is recommended that only IH "Touch-Control" or "Hydra-Touch and Touch-Control" fluid be used in the hydraulic system and the reservoir fluid level should be maintained at the "Full" mark on dip stick. Whenever the hydraulic lines have been disconnected, reconnect the lines, fill the reservoir and with the reservoir filler plug removed, cycle the system several times to bleed air from the system; then, refill reservoir to "Full" mark on dip stick and install the filler plug.

TESTING
All Models

221. The unit construction of the Hydra-Touch system permits removing and overhauling any component of the system without disassembling the others. However, before removing a suspected faulty unit, it is advisable to make a systematic check of the complete system to make certain

which unit (or units) are at fault. To make such a check, use IH test fixture No. SP-121-A (Fig. IH740) or equivalent and proceed as follows:

NOTE: High pressure (up to 1200-1500 psi) in the system is normal only when one or more of the control valves is in either lift or drop position. When the control valve (or valves) are returned to neutral, the pressure regulator valve is automatically opened and the system operating pressure is returned to a low, by-pass pressure of approximately 30-60 psi. The fact that the SE-1338-A gage, which is part of the SP-121-A test fixture, will not register this low pressure is of no importance since this low, by-pass pressure is not a factor in the test procedure.

Should improper adjustment, overload or a malfunctioning pressure regulator valve prevent the system from returning to low pressure, continued operation will cause a rapid temperature rise in the hydraulic fluid, damaging "O" rings and seals. Excessively high temperatures will cause discoloration of the paint on the hydraulic pump and manifold.

Before proceeding with the test, first make certain that the reservoir is properly filled with fluid. If not, add sufficient fluid to bring the fluid level to "Full" mark on dip stick. If necessary to add more than one quart, the pump drive shaft seal may be damaged and fluid may be leaking into the engine crankcase.

Remove the pipe plug from the regulator-safety valve block, which on series 300, 350, 350D, 400, 400D, W400, W400D, 450, 450D, W450 and W450D, requires removal of the hood left side panel, and install the SP-121-A test fixture. Hose from test fixture is connected to the reservoir filler opening. A typical test fixture installation is shown in Fig. IH741. Be sure the test

fixture shut-off valve is open; then, with all hydraulic cylinders disconnected from the system and with all of the hydraulic control valves in neutral position, start engine and run until the hydraulic fluid is at normal operating temperature.

Step 1. Advance the engine speed to ¾ throttle, move one of the Hydra-Touch control valve levers to lift position and while observing the pressure gage, rapidly close the shut-off valve until the control valve unlatches and returns to neutral position. The control valve should unlatch and return to neutral when the pressure gage reads 900-1200 psi.

NOTE: The shut-off valve must be closed with sufficient speed to give a rapid pressure rise, since a slow rise in pressure may not actuate the unlatching mechanism.

When the control valve unlatches and returns to neutral, the pressure gage should drop to its minimum reading, indicating the pressure regulator valve is functioning properly and returning the system to low pressure.

Step 2. Re-open the test fixture shut-off valve and move the same control valve lever (as used in step 1) to the drop position. Make certain, however, that the valve is set on "D" for a double acting cylinder. Refer to Fig. IH742. Rapidly close the shut-off valve until the control valve unlatches and returns to neutral. Here again, the control valve should unlatch at 900-1200 psi, from which the gage should drop to its minimum reading, indicating that the system is returned to low pressure.

Step 3. Using the same control valve as in steps 1 and 2, hold the control valve lever in the lift position, close the shut-off valve and observe the maximum pressure gage reading which should be 1200-1500 psi. Release the control valve lever and open the shut-off valve just as

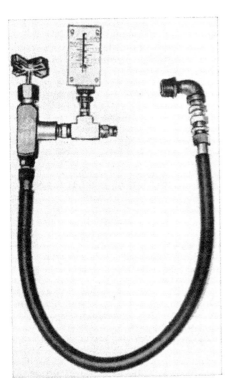

Fig. IH740—IH text fixture No. SP-121-A used for testing and trouble shooting the Hydra-Touch system.

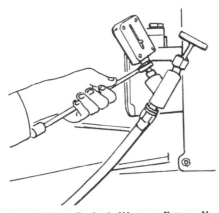

Fig. IH741—Typical IH test fixture No. SP-121-A installation. The fixture incorporates test gage SE-1338-A.

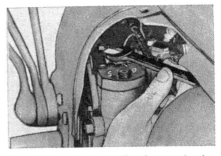

Fig. IH742—The Hydra-Touch control valve can be set (D) for double acting cylinders, (S) for single acting cylinders.

soon as the maximum pressure reading is noted. Continued operation at high pressure will cause rapid rise in the fluid temperature and damage to the system "O" rings and seals.

Step 4. If the tractor is equipped with multiple control valves, repeat steps 1, 2 and 3 for each of the other control valves.

Step 5. Stop the tractor engine and check all connections for evidence of leakage. Also check the reservoir fluid level and engine crankcase oil for evidence of internal fluid leakage around the pump shaft seal.

Step 6. Attach all hydraulic cylinders to the system and load the cylinders with their attached implements. Check each cylinder in turn as follows: With engine running at ¾ speed, control valve levers in neutral and the test fixture shut-off valve closed, operate each cylinder in turn and observe the pressure gage reading. Pressures exceeding the control valve unlatching range of 900-1200 psi indicates there is trouble in the cylinder, attaching linkage, couplings and/or attached implements.

Step 7. If the hydraulic system does not function as outlined in steps 1, 2, 3, 4, 5 or 6, refer to the trouble shooting paragraph 222 and locate points requiring further checking.

TROUBLE SHOOTING

All Models

222. The following trouble shooting chart lists troubles which may be encountered in the operation and servicing of the hydraulic power lift system. The procedure for correcting many of the causes of trouble is obvious. For those remedies which are not so obvious, refer to the appropriate subsequent paragraphs.

A. System unable to lift load, gage shows high pressure (900 psi or higher). Could be caused by:
   1. System is overloaded
   2. Damaged hydraulic cylinder
   3. Implement damaged in a manner to restrict free movement
   4. Interference restricting movement of cylinder or implement
   5. Hose couplings not completely coupled

B. System unable to lift load, gage shows little or no pressure. Could be caused by:
   1. Loss of fluid
   2. Pump failure
   3. Faulty pressure regulator valve
   4. Failure of safety valve to close

5. Leakage past cylinder piston seals
6. Plugged pump intake screen

C. System lifts load slowly, gage shows low pressure. Could be caused by:
   1. Pump failure
   2. Plugged pump intake screen
   3. Faulty pressure regulator valve
   4. Failure of safety valve to close
   5. Pressure regulator orifice enlarged or loose in block

D. With all control valves in neutral, gage shows high pressure. Could be caused by:
   1. Pressure regulator piston stuck in its bore
   2. Regulator orifice plugged

E. Loss of hydraulic fluid, no external leakage. Could be caused by:
   1. Failure of the pump drive shaft seal

F. Operating pressure exceeds 1500 psi. Could be caused by:
   1. Safety valve piston stuck in its bore
   2. Failure of safety valve spring

G. Control valve will not latch in either lift or drop position. Could be caused by:
   1. Broken garter spring
   2. Orifice in upper unlatching piston plugged
   3. Unlatching valve leakage

H. Control valve cannot be readily moved from neutral. Could be caused by:
   1. Control valve gang retaining bolts and nuts too tight, causing valves to bind in valve bodies
   2. Control valve linkage binding
   3. Orifice in upper unlatching piston plugged
   4. Scored control valve and body

I. Control valve unlatches before cylinder movement is completed, gage shows high pressure. Could be caused by:
   1. System is overloaded
   2. Damaged hydraulic cylinder
   3. Implement damaged in a manner to restrict free movement
   4. Interference restricting movement of cylinder or implement
   5. Hose couplings not completely coupled

J. Control valve unlatches before cylinder movement is completed, gage shows low pressure. Could be caused by:
   1. Weak unlatching valve spring
   2. Unlatching valve leakage

K. Control valve unlatches from lift position but not from drop position. Could be caused by:
   1. Valve set for single acting cylinder, wherein unlatching valve is inoperative in drop position
   2. Channel between upper and lower unlatching valve is plugged

L. Control valve will not center itself in neutral position. Could be caused by:
   1. Control valve gang retaining bolts and nuts too tight, causing valves to bind in valve bodies
   2. Control valve linkage binding
   3. Scored control valve and body
   4. Centering spring weak or broken
   5. Unlatching pistons restricted in movement.

M. Control valve will not automatically unlatch from either lift or drop position. Could be caused by:
   1. Loss of fluid
   2. Pump failure
   3. Faulty pressure regulator valve
   4. Plugged channels in control valve which lead to unlatching valve
   5. Leakage past the unlatching pistons
   6. Loose upper unlatching piston retainer

N. High noise level in operation of pump. Could be caused by:
   1. Insufficient fluid in reservoir
   2. Pump manifold tubes contacting some foreign part of tractor
   3. Plugged intake screen

O. Cylinder will not support load. Could be caused by:
   1. External leakage from cylinder, hoses or connections
   2. Leakage past cylinder piston rings
   3. Internal leaks in control valve

PUMP

All Models

223. An exploded view of a typical hydraulic pump is shown in Fig. IH-743. Other than installing a seal ring and gasket package, there is little actual repair work which can be accomplished on the pump.

To remove the pump, first crank engine until number one piston is coming up on compression stroke and continue cranking until last notch on crankshaft pulley is in register with pointer on crankcase front cover.

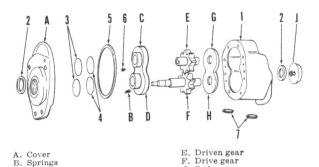

Fig. IH743 — Exploded view of a typical engine driven hydraulic pump. Numbered items show contents of the seal and gasket repair package.

A. Cover
B. Springs
C. Bearings
D. Bearings
E. Driven gear
F. Drive gear
I. Body
2. Seal
3. Fiber washer
4. Seal rings
6. Pin seal
(some models)
7. "O" rings

Remove the ignition unit and unbolt the hydraulic manifold from pump. Cover the openings in the hydraulic manifold to prevent the entrance of dirt and remove pump from engine. With pump openings covered, thoroughly wash pump in a suitable solvent to remove any accumulation of dirt. Remove nut retaining drive gear to drive shaft and using a suitable puller as shown in Fig. IH744, remove the drive gear and drive gear Woodruff key. Remove cap screws retaining cover to pump body, bump pump drive shaft on a wood block to loosen cover from pump body and remove cover. Remove seal rings (4—Fig. IH743), fiber washers (3), spring (B) and ring gasket (5). Press drive shaft seal (2) out of pump cover. Tap

drive shaft (F) on a wood block to loosen bearings (C & D) from pump body, then remove the bearings. Pin seal (6), used on some models, will come out with the bearings.

Remove the ignition unit drive lug (J) and remove gears (E & F). Tap body on wood block to remove bearings (G & H). Identify the bearings so they can be installed in their original position. Press seal (2) out of pump body.

Clean all metal parts in a suitable solvent and dry them with compressed air. If seal contacting surfaces on drive gear shaft are not perfectly smooth, polish them with fine crocus cloth and rewash the drive gear and shaft.

When reassembling, lubricate all parts with clean hydraulic fluid and use new gaskets and seals.

Install drive shaft seal (2) in body with lip toward center of pump. Install bearings (G & H) in their original position with milled slot on pressure side. Install gear (E) with tool marked side of gear toward pump body. Install gear (F), being careful not to damage seal (2) in body. A seal jumper (Fig. IH745) can be used to avoid damaging the seal. Install bearings (C and D—Fig. IH743) and on models so equipped, install pin seal (6). Notice that long bearing (D) fits over the drive shaft. Install seal ring (5) and the ten springs (B). Install seal (2) in cover so that lip of seal

faces center of pump. Install back up washer (3) and seal ring (4) in cover. Install pump cover using the seal jumper to avoid damaging the seal. Install cover cap screws and tighten them to a torque of 25 ft.-lbs. Install the ignition unit drive lug, drive shaft Woodruff key and drive gear.

When installing the pump, reverse the removal procedure and check the ignition timing.

CONTROL VALVES
All Models

224. Tractors may be equipped with either a single, double or triple valve system. All valves are identical on multiple valve systems; therefore, this section will cover the overhaul of only one valve.

The procedure for removing the 300U, 350U and 350DU control valve is evident. On other models, proceed as follows: Drain cooling system, remove the hood sections and disconnect the belt pulley control rod. Disconnect the heat indicator sending unit from cylinder head, oil pressure gage line from cylinder block and choke control at carburetor. Remove the cover from battery cables and unbolt the head light brackets from instrument panel. Unbolt and move the instrument panel rearward as far as possible. The remaining removal procedure is evident. Refer to Fig. IH746. Be sure to mark the relative position of the small control lever with respect to the

Fig. IH744—Removing drive gear from hydraulic pump. Care should be taken to avoid damaging the gear during this operation.

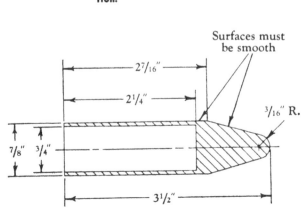

Surfaces must be smooth

2⁷/₁₆"

2¹/₄"

³/₁₆" R.

⁷/₈" ³/₄"

3¹/₂"

Fig. IH745—Home made tool which can be used to avoid damaging hydraulic pump seals when installing the pump drive shaft. Tool can be made from 7/8-inch diameter steel.

Fig. IH746 — Instrument panel assembly moved rearward in preparation for removing the Hydra-Touch system control valves on some models.

shaft (29—Fig. IH747) to insure correct assembly.

Thoroughly clean the removed control valve unit in a suitable solvent, refer to Fig. IH747 and proceed as follows:

Remove body cover (23) and gasket (26). Drift out roll pin (22) and remove indexing bushing (25). Remove roll pin (20) and withdraw lever shaft (29), bushing (19) and yoke (27). Lift out the valve guide (21) and body plug (28). Remove cap (1), garter spring sleeve (4) and garter spring (3). Push the control valve spool assembly from the valve body. Clamp the spool in a soft jawed vise and remove the upper retainer (17); then withdraw the unlatching valve (11) and its spring (12). Remove the unlatching piston (16) and guide (13) from the upper retainer. Unscrew the orifice plug (15) from piston (16). Turn the spool over in the vise and remove the lower retainer (5), retainers (6), centering spring (7) and lower unlatching piston (9).

The unlatching valve spring (12) should have a free length of 1⅝ inches and should test 12 lbs. at ⅞-inch.

The garter spring (3) should be renewed if outer diameter shows excessive wear. Original clearance of valve spool (10) in body (30) is 0.0004-0.0007. Renew the matched spool and body units if clearance is excessive or if either part shows evidence of scoring or galling. Thoroughly clean channels and the small bores in the valve spool (10). Inspect the unlatching valve (11) and its seat in the control valve spool for damage. If seat in spool (10) is destroyed, renew the spool and valve. If the seat is not damaged beyond repair, recondition the seat as follows: Place a new valve (11) on the seat and using a piece of ⅜-inch copper tubing slipped over the valve stem, rotate the valve on its seat and lightly tap the tubing with a hammer four or five times. The new valve used for reseating purposes must be discarded and a new valve used for assembly. Width of valve seat should be as narrow as possible to insure a good seal.

Clean the orifice plug and screen (15) with compressed air and make certain that bore in upper piston (16)

as well as the passages in body (30) are open and clean.

When reassembling, dip all parts in clean hydraulic fluid, then use new "O" ring seals and gaskets and reverse the disassembly procedure. Special bullet tool No. ED-3396 will facilitate installation of the garter spring (3) and spring sleeve (4). When installing guide (21), make certain that strap of guide is toward same side of body as shown in Fig. IH747. Install yoke (27) with serrated end of yoke bore toward same side of body as bushing (19). Insert shaft (29) with pin hole up; then install bushing (19) and roll pin (20). Install body cover (23) so that indexing bushing (25) engages guide (21). Install the small lever on outer serrations of shaft (29) in its original position. Note: On multiple valve systems, the relative position of the lever with respect to shaft (29) must be the same on all valves.

When installing the control valve, tighten the retaining bolts to a torque of 25 ft.-lbs. Over-tightening will result in distorted body and binding valve spool.

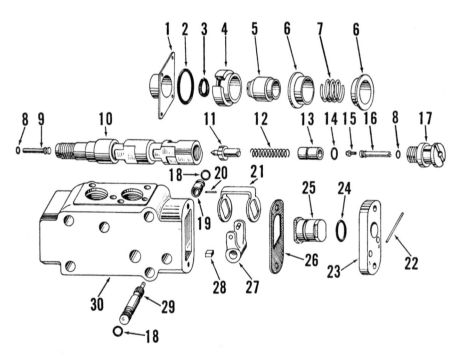

**Fig. IH747—Exploded view of a Hydra-Touch control valve. Tractors may be equipped
with a single valve system or a multiple valve system using either two or three valves.**

1. Valve cap	8. Seal ring	15. Control valve orifice plug and screen
2. Seal ring	9. Lower unlatching piston	16. Upper unlatching piston
3. Garter spring	10. Control valve spool	17. Upper unlatching piston retainer
4. Garter spring sleeve	11. Unlatching valve	18. Seal ring
5. Lower unlatching piston retainer	12. Unlatching valve spring	19. Bushing
6. Control valve centering spring retainer	13. Unlatching piston guide	20. Roll pin
7. Control valve centering spring	14. Seal ring	21. Guide

22. Roll pin
23. Body cover
24. Seal ring
25. Indexing bushing
26. Gasket
27. Yoke
28. Body plug
29. Control valve lever shaft
30. Control valve body

REGULATOR AND SAFETY VALVE BLOCK

All Models

225. On models with power steering, the regulator and safety valve block is fitted with a flow control valve composed of items (16 through 26—Fig. IH748). On models without power steering, the unit is similar except the flow control valve is not used.

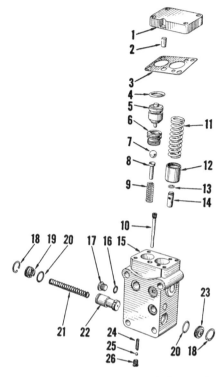

Fig. IH748—Exploded view of the regulator and safety valve unit used on models with power steering. The unit used on models without power steering is similar except it is not equipped with flow control valve (items 16 through 26).

1. Housing cover
2. Dowel pin
3. Gasket
4. Seal ring
5. Regulator valve piston
6. Regulator valve seat
7. Ball
8. Ball rider
9. Regulator valve ball rider spring
10. Safety valve orifice screen and plug
11. Safety valve spring
12. Safety valve spring retainer
13. Snap ring
14. Safety valve piston
15. Housing
16. Seal ring
17. Plug
18. Snap ring
19. Flow control valve spring retainer
20. Seal ring
21. Flow control valve spring
22. Flow control valve
23. Flow control valve retainer
24. Auxiliary safety valve spring
25. Ball
26. Plug

The procedure for removing the 300U, 350U and 350DU valve block is evident. On other models, the hood and fuel tank must first be removed.

Thoroughly clean the removed control valve unit in a suitable solvent, refer to Fig. IH748 and proceed as follows:

Remove two diagonally opposite cap screws retaining the cover (1—Fig. IH748) to the body and insert in their place 2-inch long cap screws and tighten them finger tight. Remove the two remaining short cap screws; then relieve the pressure of the safety valve spring gradually by alternately unscrewing the 2-inch long screws. Remove cover (1) and gasket (3). Withdraw the safety valve spring (11), spring retainer (12) and safety valve (14). Remove the regulator valve piston (5), unscrew the valve seat (6) and remove the ball (7), rider (8) and spring (9). Unscrew and remove the orifice plug and screen (10). On models with power steering, remove snap rings (18) and withdraw the flow control valve spring and valve. Remove the auxiliary safety valve (24, 25 and 26).

Inspect ball valve (7) and valve seat (6) for damaged seating surfaces. Piston (5) and safety valve (14) must be free of nicks or burrs and must not bind in the block bores. Clean the orifice plug and screen (10) with compressed air and make certain that bores in block (15) are open and clean.

When reassembling, dip all parts in clean hydraulic fluid, use new "O" ring seals and gaskets and reverse the disassembly procedure. Be sure that all openings in block, gasket (3) and cover (1) are aligned.

CHECK VALVE AND DROP RETARDING VALVE

Models So Equipped

226. The procedure for removing, overhauling and/or cleaning the check valve and the drop retarding valve is evident after an examination of the unit and reference to Figs. IH749 and 750.

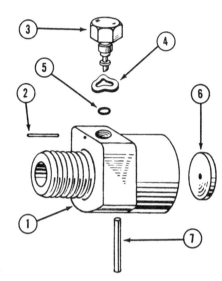

Fig. IH750—Exploded view of the drop retarding valve used on some models.

1. Body
2. Roll pin
3. Regulator knob
4. Tension spring
5. Seal ring
6. Washer
7. Taper pin

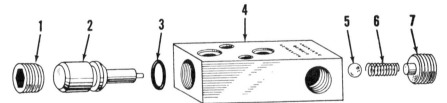

Fig. IH749—Exploded view of the Hydra-Touch system check valve block used on some models.

1. Plug
2. Piston
3. Seal ring
4. Block
5. Ball
6. Check valve spring
7. Plug

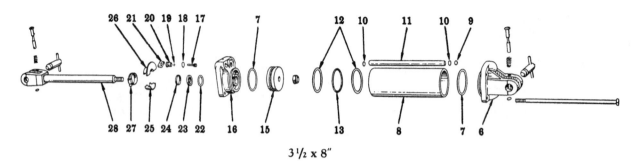

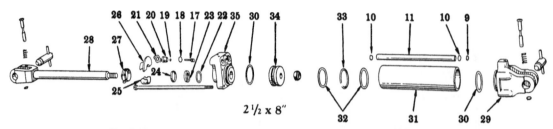

Fig. IH751—Exploded view of the two sizes of hydraulic cylinders.

6. Head	12. Back-up washer	19. Seal ring	25. Collar half	31. Cylinder tube
7. Seal ring	13. Seal ring	20. Retainer	26. Collar half	32. Back-up washer
8. Cylinder tube	15. Piston	21. Retainer plate	27. Clamp	33. Seal ring
9. Lock washer	16. Head	22. Gland ring	28. Piston rod	34. Piston
10. Seal ring	17. Limit stop valve	23. Dust seal	29. Head	35. Head
11. Return tube	18. Seal ring	24. Dust seal retainer	30. Seal ring	

NOTES